PROCEEDINGS OF THE 1ST INTERNATIONAL CONFERENCE ON HALAL DEVELOPMENT (ICHAD 2020), MALANG, INDONESIA, 8 OCTOBER 2020

# Halal Development: Trends, Opportunities and Challenges

*Edited by*

Heri Pratikto & Ahmad Taufiq
*Universitas Negeri Malang, Indonesia*

Adam Voak
*Deakin University, Australia*

Nurdeng Deuraseh
*Universitas Islam Sultan Sharif Ali, Brunei Darussalam*

Hadi Nur
*Universiti Teknologi Malaysia, Indonesia*

Winai Dahlan
*The Halal Science Center Chulalongkorn University, Bangkok, Thailand*

Idris & Agus Purnomo
*Universitas Negeri Malang, Indonesia*

**CRC Press**
Taylor & Francis Group
Boca Raton London New York Leiden

CRC Press is an imprint of the
Taylor & Francis Group, an **informa** business
A BALKEMA BOOK

*CRC Press/Balkema is an imprint of the Taylor & Francis Group, an informa business*

© 2021 The Author(s)

Typeset by MPS Limited, Chennai, India

The right of the first International Conference on Halal Development (ICHaD2020)
to be identified as author[/s] of this work has been asserted by him/her/them in
accordance with sections 77 and 78 of the Copyright, Designs and Patents Act 1988.

*Library of Congress Cataloging-in-Publication Data*
A catalog record has been requested for this book

Published by:   CRC Press/Balkema
                Schipholweg 107C, 2316 XC Leiden, The Netherlands
                e-mail: enquiries@taylorandfrancis.com
                www.routledge.com – www.taylorandfrancis.com

ISBN: 978-1-032-03830-8 (Hbk)
ISBN: 978-1-003-18928-2 (eBook)
ISBN: 978-1-032-03835-3 (Pbk)
DOI: 10.1201/9781003189282

# Table of contents

# Preface

The increasing demand for halal products including goods and services every year, especially for food and beverages, has resulted in a growing need for products with Halal guarantees. Along with the increasing trend of demand globally, it has resulted in an increase in producers of halal food and beverages in both Muslim and non-Muslim countries. The growth in demand for halal products increased significantly in Asia, Africa and Europe. This has also resulted in an increase in demand from consumers of halal products so that producers in producing food and beverage products must comply with the provisions of Islamic law.

In addition, not only for food and beverage products but also the demand for halal tourism is also increasing. Indonesia is one of the largest Muslim countries in the world. However, there are still many Muslim consumer actors and Muslim producer actors who do not yet have an awareness of the importance of complying with the provisions of Islamic law in consuming and producing goods and services. There are still many restaurants and hotels that serve food and drinks that are not certified halal. There are still many food, medicinal and cosmetic products that are not halal certified. But now many secular countries such as France, Canada, Australia, the United States, Britain are also halal certified with the aim of meeting the Muslim demand for halal products for food and beverage, including for halal tourism.

The first International Conference on Halal Development (ICHaD) 2020 was held at Universitas Negeri Malang, East Java-Indonesia on October 8, 2020. This Conference was organized by Halal Center, Institute for Research and Service of Engagement Universitas Negeri Malang. There were nine main speakers in this conference including Prof. Ir. Sukoso, M. Sc., Ph.D, from The Organizing Agency for the Halal Product Guarantee (BPJPH), Ministry of Religion Affairs of the Republic of Indonesia, Adam Voak, Ph. D., from Institute of Supply Chain and Logistics (CSCL) Deakin University Australia, Prof. Dr. Irwandi Jaswir, M.Sc. from International Islamic University Malaysia (IIUM), Sycikh Dr. Salim Alwan Al Husaini from Darul Fatwa Australia, Dr. Mohd Iskandar Bin Illyas Tan, from Universiti Teknologi Malaysia (UTM), Prof. Dr. Nurdeng Deuraseh, Ph.D from University Islam Sultan Sharif Ali (UNISSA) Brunei Darussalam, Prof. Dr. Winai Dahlan from Halal Science Center Chulalongkorn University Thailand, Prof. Dr. Shuhaimi Mustafa from University Putra Malaysia (UPM), and Prof. Dr. Suhaimi Ab Rahman from University Putra Malaysia (UPM).

The organizers wish to acknowledge the keynote speakers for their presentation on ICHaD 2020. We also acknowledge publicly the valuable services provided by the reviewers for their time, hard work, and dedication to this Conference. In addition, many thanks given for all persons who help and support this conference. Furthermore, we also invite the presenters around the world to participate in the 2nd ICHaD that will be held in 2021. Finally, we hope that the future event will be as successful as indicated in this proceeding.

# Acknowledgement

The editor and organizer of the Conference on Halal Development (ICHaD) 2020 wishes to acknowledge the keynote speakers for their presentation on ICHaD 2020. They are Adam Voak, Deakin University Australia; Assoc. Prof. Dr. Winai Dahlan, HSC-Chulalongkorn University, Thailand; Prof. Suhaimi Ab Rahman, UPM Malaysia; Dr. Mohd Iskandar Bin Illyas Tan, Universitas Teknologi Malaysia (UTM) Malaysia; Prof. Sheikh Salim Alwan Al Hussainy, Darulfatwa, Sydney Australia; Prof. Dr. Irwandi Jaswir, M.Sc, Universitas Islam International Malaysia (UIIM); Associate Prof. Dr. Nurdeng Deuraseh, University Islam Sultan Sharif Ali (NISSA) Brunei Darussalam; Prof. Ir. Sukoso, M.Sc., Ph.D, Head of Halal Product Guarantee Agency (BPJPH) Republic of Indonesia.

The organizers also wish to acknowledge publicly the valuable services provided by the reviewers. On behalf of the editors, organizers, authors and readers of this Conference, we wish to thank the keynote speakers and the reviewers for their time, hard work, and dedication to this Conference. Without their services, the editors could not maintain the high standards of Halal Development Research.

The organizers wish to acknowledge Prof. Hadi Nur, Ph.D, Prof. Dr. Markus Diantoro, M.Si, Dr. Achmad Munjin Nasih, Ani Wilujeng, Ph.D, Dr. Achmad Taufiq for the discussion, suggestions, and cooperation to organize the keynote speakers of this Conference. The organizers also wish to acknowledge speakers and participants who attended this seminar. Many thanks given for all persons who helped and supported this conference.

The organizers wish to apologize to the speakers who could not publish his/her paper in this Conference Proceeding. Our apology also given to all participants for all shortcomings in this conference. See you in the next ICHaD 2021 at the Universitas Negeri Malang.

Malang, December 15, 2020
Local Organizer of ICHaD 2020
The Halal Centre, Institute for Research and Community Engagement (IRCE),
Universitas Negeri Malang

# Scientific committee

Prof. Dr. Heri Pratikto, M.Si.
*Universitas Negeri Malang, Indonesia*

Prof. Dr. H. Budi Eko Soetjipto, M.Ed., M.Si
*Universitas Negeri Malang, Indonesia*

Dr. Agus Hermawan, GradDipMgt., M.Si, M.bus
*Universitas Negeri Malang, Indonesia*

Dr. Puji Handayati, S.E., M.M. Ak., CA, CMA, CIBA, CSRS, CSRA
*Universitas Negeri Malang, Indonesia*

Dr. Sri Pujiningsih, S.E., M.Si., Ak
*Universitas Negeri Malang, Indonesia*

Dr. Grisvia Agustin, S.E., M.Sc.
*Universitas Negeri Malang, Indonesia*

Dr. Dra. Asfi Manzilati, ME.
*Universitas Brawijaya, Indonesia*

Dr. H. Ahmad Djalaluddin, Lc., M.A.
*UIN Maulana Malik Ibrahim Malang, Indonesia*

Akhmad Akbar Susamto, S.E., M.Phil., Ph.D.
*Universitas Gajah Mada, Indonesia*

Dr. Rosmiza Bidin
*Universiti Putra Malaysia, Malaysia*

A.P. Dr Rodziah Atan
*Universiti Putra Malaysia, Malaysia*

A.P. Dr Nangkula Utaberta
*Universiti Putra Malaysia, Malaysia*

Assoc. Prof. Dr. Winai Dahlan
*Chulalongkorn University, Thailand*

Prof. Madya Dr. Nurdeng Deuraseh
*University Islam Sultan Sharif Ali, Brunei Darussalam*

# Organizing committee

**Advisory Board**
Prof. Dr. Ah. Rofi'uddin
*Universitas Negeri Malang, Indonesia*

**Chairman:**
Prof. Dr. Heri Pratikto, M.Si.
*Universitas Negeri Malang, Indonesia*

**Secretary**
Assoc. Prof. Dr. Sukarni
*Universitas Negeri Malang, Indonesia*

**Treasurer**
Assoc. Prof. Dr. Irma Kartika Kusumaningrum
*Universitas Negeri Malang, Indonesia*

**Technical Chair**
Andro Agil Nur Rakhmad, S.E.I., M.E.
*Universitas Negeri Malang, Indonesia*

**Publication**
Dr. Ahmad Taufiq, S. Pd., M. Si
Asst. Prof. Aripriharta, Ph.D
Poppy Puspitasari, Ph.D
Idris, S.S., M.M.

# Building human capability to ensure halal supply chain compliance

A. Voak*
*Deakin University, Melbourne, Australia*

B. Fairman
*ASEAN Institute of Applied Learning, Indonesia*

ABSTRACT:   Halal supply chain compliance is fundamentally built on trust. A trust that all supply chain actors will ensure the hygiene, cleanliness, safety and wholesomeness of the product and the services they provide from origin to consumption. Halal supply chain compliance is not just the responsibility of regulators, certifiers and other quality assurance stakeholders. It involves a fundamental commitment to managing the product, communication and financial practices within the supply chain so that they comply with Shariah Law. With the making of this commitment come unique Human Resource Development (HRD) challenges and opportunities to ensure all supply chain actors have the skills, knowledge and attitudes to ensure that trust and respect are maintained. This paper focuses on the human capability development interventions needed to ensure Halal supply chain compliance and to assure brand claims remain consistent with the expectations of Muslim and non-Muslim consumers. The growing complexity of global supply chains, along with the interdependency of the actors involved at each stage, means that considerable effort and investment should be made on better understanding the management human actors within this process. The people component is a fundamental piece of the compliance puzzle, in not only meeting the regulatory and certification requirements of Halal but also establishing the respectful practice of all stakeholders needed to maintain trust within the system.

*Keywords*:   human capability development, halal supply chain, halal logistics

## 1   INTRODUCTION

People fundamentally drive halal supply chains and its service quality is largely dependent on the nature of personal inputs and outputs. In a Halal supply chain, the sum of these individual activities must combine seamlessly across all business activities to ensure Sharia law compliance from origin to consumption (Omar & Jaafar 2011). However, global trade complexity and the lengthening of food supply chains have made compliance increasingly more challenging. It has become evident that with this increase in complexity comes an urgent need for increased human resource development. This human resource development must focus on new levels of knowledge, skills, abilities, values and social assets the employee must possess which will ultimately contribute to the excellence of their performance (Hashim & Shariff 2016) and consequently to the assurance of strict compliance. This Halal compliance extends well beyond what is considered permissible, acceptable or indeed lawful and includes standards, guidelines and processes related to hygiene, cleanliness, quality, trustworthiness and safety (Ab Talib et al. 2015; Supian et al. 2019).

Modern Halal supply chains are characterized and shaped by their interdependencies, which are tested as products move through every stage of production and between vast distribution networks towards the final consumer (Rejeb 2018). Globalization has illuminated a range of challenges and

---

*Corresponding author

risks for Halal food supply chains (Kamalahmadi & Parast 2016; Supian et al. 2019). Globalization has necessitated that food travels greater distances (Ali et al. 2017) and the longer the chain, the greater its vulnerability (Supian et al. 2019). The strength of the links between supplier efficiency, supply chain performance and customer service has now become, in a significant way, critical to the maintenance of trust and respect of services.

To ensure their long-term viability, suppliers and producers now need to improve their association with retailers and customers to ensure they meet heightened expectations. These growing customer demands mean more emphasis is being placed on efficiencies and speed to market. While efficiency and effectiveness are fundamental principles of any supply chain, suppliers must also be mindful of product quality maintenance and the trust placed in certification regimes, all of which can impact on Halal integrity (Yaacob et al. 2016). As consumers become better informed and knowledgeable, and as global food safety crises and incidents sharpen their awareness, they become more concerned with food origin, authenticity, safety and quality (Abd Kadir et al. 2016). Therefore, the human element becomes a critical piece of the puzzle, as workforces build the resultant capability needed along Halal food supply chains to respond to these market expectations.

## 2 HALAL FOOD SUPPLY CHAIN COMPLIANCE

Compliance is becoming more complex, with Halal supply chain producers and their service-related actors required to ensure transparent information flows which outline not only origin and composition but also their certification or regulatory status. Further, many organizations have embraced global sourcing to reduce costs, with this often resulting in unintended risks and unanticipated consequences. These practices have introduced further complexity by creating interdependency amongst and between supply chain actors, all of which must meet Halal requirements. To better understand these interdependencies, interrelationships and resultant compliance risks, the mapping of the Halal supply chain is crucial. Mapping Halal supply chains creates greater visibility by identifying and mitigating non-compliance and critical control points which need to be effectively monitored.

There are many hygiene, cleanliness, safety and wholesomeness requirements that should be controlled, practiced and followed to avoid disruptions to the Halal supply chain. Potential regulatory scrutiny, for example, could lead to negative impacts on the organization's bottom line and reputation. Therefore, a Halal supply chain compliance program must serve as an enabler to not only better control product quality, transparency of information and regulated financial flows, but also to increase their visibility. A compliance program is thus a vital element of a broader enterprise risk mitigation strategy. To mitigate the risk of non-compliance, suppliers must try to understand the linkages and interdependencies that exist between supply chain actors more clearly. Comprehensive enterprise Halal supply chain risk management must implement the diligent anticipation, monitoring, management and assurance of operational practices of all actors. A harmonized and integrated Halal supply chain compliance program must include appropriate risk-management activities and controls through all three major supply chain flows, those of financial, product and information flows from producer to consumer.

## 3 FACETS OF COMPLIANCE AND HUMAN CAPABILITY REQUIREMENTS

### 3.1 *Segregation, cross contamination, food hygiene and cleaning*

Currently, the main emphasis of Halal compliance is being placed on the assurance, regulation and certification of food production processes. These associated standards aim to control the cleanliness, hygiene and cross-contamination with non-Halal foods while in production, manufacturing, on-farm facilities and in slaughtering facilities. Physical segregation is not only critical in the production phase but also along the supply chain to preserve its Halal status (Abd Kadir et al. 2016).

Suppliers and producers must, therefore, ensure that logistics equipment, facilities and the associated infrastructure that support distribution meets compliance requirements. This infrastructure and equipment must be consistently monitored, maintained, cleaned and the necessary mechanisms implemented to avoid contamination. To ensure complete chain compliance, suppliers must ensure that they have controlled all stages of distribution, including transparency in product movement. This transparency or visibility is vital in securing those logistics-related services which are needed to deploy preventive or mitigating measures along the supply chain.

All Halal supply chain actors must ensure appropriate hygiene maintenance is undertaken. This program should include regular vetting of cleaning, sanitation, waste and pest control systems of all logistics service providers. Established protocols must be followed, and personal hygiene of distribution staff must be monitored. These compliance regimes should also extend to adequate hygiene control measures and protocols while product is in transit. Further, Halal products must be clearly labelled and visible for identification purposes. Additionally, all workers, performing operational duties should be adequately trained in Halal food hygiene and protocols to maintain continuity and integrity of product flows (Nor et al. 2016).

### 3.2 Halal provenance, fraud and traceability

With the increase in complexity of food supply chains, consumer concerns over the provenance of Halal food that is claimed to have been prepared in accord with religious requirements, are gaining importance. Instances of improper identification and sale of Halal meat-based products, for example, have given rise to questions over the authenticity of such foods (Thomas et al. 2017). Halal food fraud appears to be on the rise since an increasing number of cases regarding the misuse of Halal Certification and Regulatory logos have been identified. These fraudulent practices indicate that there are weaknesses in unequivocally determining product compliance (Supian et al. 2019). Halal food traceability is a shared responsibility of all actors and stakeholders along the Halal supply chain (Supian et al. 2019). Halal traceability aims to identify the product, process, participant and marketing attributes and associated logistics information in the supply chain seen from an upstream as well as from a downstream perspective, and handling should be recorded at each node and stored centrally in a secure database (Khan et al. 2018).

To further facilitate traceability implementation, it is incumbent on Halal certification and regulatory bodies to introduce standardization protocols using the same accreditation standards (Hassan et al. 2016). Traceability and tracking systems are tools for rapid communication, and aim to increase the transparency of information flows within the supply chain (Zailani et al. 2010), all of which require significant collaboration among all the parties involved (Mohd Saifudin et al. 2017). It is of note that a considerable hurdle for the continued security and resultant growth of the Halal industry is the lack of agreed Halal standards that are acceptable to Muslims worldwide; it is a matter of record that each country has its own Halal compliance standards which make control and checking difficult (Majid et al. 2015).

## 4 DEVELOPING HALAL SUPPLY CHAIN HUMAN CAPABILITY

Job knowledge (the how) and Shariah knowledge (the why) are required for developing human capability along the Halal supply chain. Not only do actors along the supply chain require the job skills and knowledge relevant to their traditional operational roles, but within a Halal environment, they will also need to acquire new knowledge. This knowledge acquisition should target Halal Islamic rulings and Shariah law, awareness of Halal-Haram principles and appropriate Halal management practices (Shariff et al. 2016). To ensure Halal compliance, professional standards of practice and ethics for every key role along the chain, should be agreed upon and maintained, through widely recognized National and International professional associations (Shariff et al. 2016).

Previous research has identified critical issues regarding Halal logistics, focused on the governance of practices, these include: (i) the transportation and distribution security of food, (ii) the regulated storage and warehousing of perishable goods, and (iii) the manner of materials handling and operational processing at terminal nodes (Tieman et al. 2012; Tieman 2013; Talib et al. 2014; Ab Talib et al. 2015). To meet these Halal operational requirements, each of these facets must ensure the physical separation of Haram (non-Halal) products, even if they are inside containers or on different shelves in the cold-rooms or warehouses.

The requirements for keeping track of changes in technology and the related challenges of information management are now commonplace in supply chains. Employees must be adequately trained to allow them to access and use technology related to the transport and provenance of material under their control. It is well recognized that non-stop technological change adds to the many other human resources challenges which are related to job redesigning and the development of new skill sets. These concerns overlap with Halal training programs in that technological skills are very much needed to provide transparent Halal management. Sharia knowledge is also clearly essential if people within the supply chain are to be able to interpret food handling conditions appropriately. However, what currently makes this situation somewhat complicated, is that consistency of existing programs of instruction for personnel regarding transport duration, protection of contents and coverage of emergencies (Jamaluddin et al. 2015) are not built on a foundation of agreed occupational standards. Whilst training is deemed necessary by supply chain employers; this requires financial investment and other opportunity costs for their employees. An adequate "training needs" analysis along the supply chain needs to be conducted to ensure any human capital interventions close competency halal compliance knowledge, skills, and attitude gaps (Majid et al. 2019). Against this uncertain background, there is also a real need to develop skill sets and micro-learning opportunities to manage fast-changing technologies which will support Halal compliance.

## 5  HRD OPPORTUNITIES AND HALAL LOGISTICS HUBS

The task facing HRD professionals is a complex one since the set of requirements imposed by Halal authorities is difficult to integrate within the complex and intricate transactions common to supply chain soft and hard infrastructure. Issues of planning, collaboration, teamwork, leadership and creativity, which are part of a growing supply chain structures, must now be cast against often varying cross-border sets of compliance principles.

In this in-between period, where supply chain actors and Halal requirements are becoming more widely acknowledged, there is a need for the sharing of knowledge and techniques within the area. A possible vector here could be the use of 'Communities of Practice' where ideas and practices can be detailed and considered by a range of stakeholders. One such community of practice could involve a careful investigation of logistics hubs and transport modality integration. The notion of Halal logistics integration and coordination could potentially contribute to greater compliance and the assurance of products (Karia 2019).

Findings from a desk review of current Halal compliance practices and approaches have highlighted that logistics systems face significant challenges in keeping pace with the global growth of the sector. Further, the human actors involved in logistics-related services and activities require significantly more training, particularly around Halal awareness and the resultant operational compliance protocols. There is an urgent requirement for more reliable and standardized human resource development systems which clearly enunciate the skills, knowledge and attitudes needed by Halal logistics personnel. This integrated planning would set the scene for processes which could be reflected in agreed cross-border Halal occupational standards (Voak & Fairman 2020). The review also revealed that greater clarity is needed around the competencies needed from producer to consumer to manage Halal supply chain compliance effectively. With these lessons in mind, more significant effort needs to be placed on raising the importance of the human element and its impact on Halal compliance. The subsequent human resource development needs for this sector will need to consider and determine how HR for Halal supply chain compliance can be strengthened particularly with a focus on the potential of Halal logistics hubs (Mahidin et al. 2016).

# 6  CONCLUSION

Apart from the fundamentals of supply chain management, more significant cooperation between supply chain actors and other disciplines are critical in provisioning Halal food compliance and their resultant integrity. Human resource planning, management and development along the whole Halal supply chain must be comprehensively and systematically addressed to provide consumer trust. Halal supply chain compliance is only as strong as the human beings that serve the food ecosystem. For human capability development to accord with Halal compliance, it must go beyond current knowledge and skills development. Human development interventions along the chain must look closely at building the cultural competence, attitudes and values necessary to prevent fraud, preserve food integrity and the integrity of Halal regulatory and certification systems.

The above discussions illustrate some of the complexities that organizations face in designing Halal supply chain compliance programs, and the measures required to mitigate risks. Further research is needed to develop a fuller understanding of the appropriate and responsive occupational standards that may be necessary to assist organizations in responding to the growing complexities of Halal supply chain compliance. Such an investment would assist in the minimizing of disruptions and breaches that could impact on Halal brand reputation and consumer trust.

# REFERENCES

Ab Talib, M. S., Hamid, A. B. A., & Zulfakar, M. H. (2015). Halal supply chain critical success factors: a literature review. *Journal of Islamic Marketing*.

Abd Kadir, M. H., Rasi, R. Z. R. M., Omar, S. S., & Manap, Z. I. A. (2016). Halal Supply Chain Management Streamlined Practices: Issues and Challenges. IOP Conference Series: Materials Science and Engineering,

Ali, M. H., Zhan, Y., Alam, S. S., Tse, Y. K., & Tan, K. H. (2017). Food supply chain integrity: the need to go beyond certification. *Industrial management & data systems*.

Hashim, H. I. C., & Shariff, S. M. M. (2016). Halal supply chain management training: issues and challenges. *Procedia Economics and Finance*, 37, 33–38.

Hassan, W. A. W., Ahmad, R., Hamid, A., Megat, N., & Zainuddin, M. (2016). The Perception on Halal Supply Chain Management Systems Implementation of SMEs in Selangor. *Indian Journal of Science and Technology*, 9, 34.

Karia, N. (2019). Halal logistics: practices, integration and performance of logistics service providers. *Journal of Islamic Marketing*.

Khan, S., Haleem, A., Khan, M. I., Abidi, M. H., & Al-Ahmari, A. (2018). Implementing traceability systems in specific supply chain management (SCM) through critical success factors (CSFs). *Sustainability*, 10(1), 204.

Mahidin, N., Othman, S. N., & Saifudin, A. M. (2016). Halal logistics issues among the food industry companies: A preliminary study. *Journal of Global Business and Social Entrepreneurship (GBSE)*, 2(1), 34–40.

Majid, M. A. A., Abidin, I. H. Z., Majid, H., & Chik, C. T. (2015). Issues of halal food implementation in Malaysia. *Journal of Applied Environmental and Biological Sciences*, 5(6), 50–56.

Majid, Z. A., Kamarulzaman, N. H., Rahman, A., Jaafar, H. S., Rahman, N. A. A., & Mohammad, M. F. (2019). Halal Integrity from Logistics Service Provider Perspective. *Int. J Sup. Chain. Mgt*, 8(5).

Mohd Saifudin, A., Othman, S. N., & Mohamed Elias, E. (2017). Exploring in setting a model for Islamic supply chain in Malaysia. *International Review of Management and Marketing*, 7(1), 95–102.

Nor, M. R. M., Latif, K., Ismail, M. N., & Nor, M. N. M. (2016). Critical success factors of halal supply chain management from the perspective of malaysian halal food manufacturers. *Nigerian Chapter of Arabian Journal of Business and Management Review*, 62(3804), 1–23.

Omar, E. N., & Jaafar, H. S. (2011). Halal Supply Chain in the Food Industry: A Conceptual Framework. IEEE Symposium on Business, Engineering and Industrial Application (ISBEIA), Langkawi, Malaysia, Rejeb, A. (2018). Halal Meat Supply Chain Traceability Based on HACCP, Blockchain and Internet of Things. *Acta Technica Jaurinensis*, 11(1).

Shariff, S. M., Mohamad, S., & Hashim, H. I. C. (2016). Human Capital Development in Halal Logistics: Halal Professionals or Halal Competent Persons. *Journal of Applied Environmental and Biological Sciences*, 6, 1–9.

Supian, K., Abdullah, M., & Ab Rashid, N. (2019). Halal Practices Integrity and Performance Relationship: Are Halal Supply Chain Trust and Commitment the Missing Links? *Int. J Sup. Chain. Mgt Vol*, 8(6), 1045.

Thomas, A. M., White, G. R., Plant, E., & Zhou, P. (2017). Challenges and practices in Halal meat preparation: a case study investigation of a UK slaughterhouse. *Total Quality Management & Business Excellence*, 28(1-2), 12–31.

Voak, A., & Fairman, B. (2020). Anticipating Human Resource Development Challenges and Opportunities in 'Halal Supply Chains' and 'Halal Logistics' within ASEAN. *Nusantara Halal Journal*, 1(1), 1–9.

Yaacob, T. Z., Jaafar, H. S., & Rahman, F. A. (2016). An Overview of Halal Food Product Contamination Risks During Transportation. *Science International*, 28(3).

*Halal Development: Trends, Opportunities and Challenges – Pratikto et al (eds)*
*© 2021 The Author(s), ISBN 978-1-032-03830-8*

# Halal destinations in Asia: A SWOT analysis

A. Tohe*, H. Pratikto & Kholisin
*Universitas Negeri Malang, Malang, Indonesia*

R.B. Atan
*Universiti Putra Malaysia*

ABSTRACT:   This research is aimed at investigating the current state of halal destinations in four Asian countries: Indonesia, Malaysia, Thailand, and Japan. Assessment of destinations in each country was conducted using a SWOT analysis, before setting up a comparison to identify their competitive advantages and determining the relative correlation between the perceived readiness of destinations with the inbound tourist arrival. Features of assessment were based on the conceptual framework developed by COMCEC 2016, known as Muslim-Friendly Tourism (MFT), consisting of Six Faith-based Needs, Demand Side Key Themes (reasons and motivation for travel) and Supply Side Key Themes (travel & hospitality services and facilities). The results showed that three variables had been important in shaping the decision that potential inbound Muslim tourists made to visit a halal destination: religious and cultural affinity, affordability and visa-freedom. The more visible the correlation that a destination had with these three, the more likely that it received more visits. Furthermore, the first two factors were found to compensate for the alleged weaknesses found in a halal destination, as long as it provided basic halal needs.

*Keywords*:   halal tourism, Asian countries, MFT, SWOT Analysis

## 1   INTRODUCTION

Halal tourism is a new phenomenon expanding from the ever-growing halal industry (Samori et al. 2015). The increasing demand for tourism among Muslims has run parallel with the growth of the Muslim population worldwide (Henderson 2016), in addition to the improved affluence that allowed for greater consumption and expenditure (Fischer 2011). Catering to the needs of Muslims, halal tourism is highly lucrative, showing an annual growth of 5%, slightly exceeding the growth of general tourism at 4%. Predicted to be the largest religious population in some coming decades, it is the Muslims who primarily funded and moved this new model of business (Pietro & Secinaro 2019). Furthermore, halal tourism had demonstrated the propensity to expand over time. As such, halal tourism had inspired many service providers, in both Muslim and non-Muslim countries, to build new hotels, open up new attractions, start new restaurants, and offer new travel packages (Muhamad et al. 2019).

Halal tourism is growing because, in addition to Muslim countries such as Indonesia and Malaysia, non-Muslim countries have also participated in providing halal services and destinations (Muhamad ct al. 2019). In Asia, countries such as Japan and Thailand had taken advantage of this opportunity by offering Muslim-friendly facilities and services to inbound Muslim visitors. Halal food was available in many places in the two countries: on flight, in the airports, or some restaurants downtown, just as prayer facilities were more easily found in public places, although not adequately enough to satisfy the needs of the Muslim tourists visiting (Hamid et al. 2018).

---

*Corresponding author

DOI 10.1201/9781003189282-2

Tourism had contributed quite significantly to Indonesia's economic growth (Jaelani 2017), especially when the national economy was facing a global crisis by showing an increase in the terms of the export of goods and services (Widagdyo 2015). GMTI (2019) ranked Indonesia as the most favorable halal destination among members of the Organization of Islamic Cooperation (OIC) countries. Indonesia had the best opportunity to lead the halal tourism industry because it might be the country best able to provide what the Muslim tourists need, not only in terms of practicing their religion while traveling, but also in terms of a sense of security, safety, and familiarity. For Middle Eastern Muslim tourists, in particular, Indonesia could offer rich natural and cultural tourism sites that are markedly different from the ones found in their place of origin (Mubarok & Imam 2020). The Muslim-friendly ambience that Indonesia offered, due to the religious and cultural affinity and the wide availability of halal services and facilities, were the most attractive factors for inbound Muslim tourists to visit (OIC 2017), in addition to its affordability that potentially improved tourists' purchasing power.

Furthermore, the government's strong commitment to support halal tourism in the country had been well attested to (OIC 2017). In 2015, the Ministry of Tourism and Creative Economy of Indonesia launched "Muslim Guides," in addition to forming a special committee that oversaw the development of halal tourism and its promotion. The government had also organized several campaigns by orchestrating social media influencers to portray Indonesia as a Muslim-friendly destination. It had chosen, based on 80 parameters, some tourism regions—namely Lombok, West Sumatra, and Aceh—as competitive destinations (OIC 2017). Moreover, Indonesia had been actively involved in some international halal tourism events and forums. These and other governmental efforts had made Indonesia the most favorable destination (GMTI 2019).

Malaysia has been consistently one of the top ranked destinations worldwide in halal tourism. The Master Card Crescent Rating had positioned Malaysia on top among OIC members in terms of the friendliness of its halal facilities and services (Hamid et al. 2018). GMTI 2019 also put Malaysia, along with Indonesia, as the most favorable halal destinations among OIC countries. Not only had tourism been significant in its economy, but it had also demonstrated a constant increase in supporting Malaysia's state revenue over time, in addition to providing opportunities for new jobs (Hamid et al. 2018).

Similar to Indonesia, Malaysia had benefited from its status as a Muslim majority country, despite some deficiencies in its handling of halal tourism. While tourism in Malaysia was more Western-inspired and did not pay special attention to Muslim tourists' needs, but its status as a Muslim country had created some guarantee for Muslim travelers that they would not have any difficulties in finding Muslim-friendly facilities during their visit. However, on a larger scale, in places where tourists come in huge groups, this lack of accommodation could present a challenge to cater to the needs of tourists, especially in regard to halal food (Hamid et al. 2018).

In the spirit of capturing the booming market, Japan had responded to the needs of Muslim tourists by training hospitality staff through a "Muslim-Friendly Project" (Battour 2018), which aimed at increasing staff's understanding of Muslims' needs and concerns. All efforts that Japan made to restructure its tourism to be more Muslim-friendly had made the country as one of the most popular destinations for Muslim travelers (Hariani 2017). Of non-OIC countries, Japan ranked second as a favored destination (GMTI 2019), showing some improvement from 2017 and even replacing Thailand that had been in that second position in previous years. Japan witnessed an almost double increase of inbound Muslim visitors between 2010 and 2015 (Henderson 2016). Such a significant increase could have been the result of a combination of factors, such as the ease of air connectivity and visa requirements, in addition to intensive campaigns and the devaluation of the Yen that allowed for more affordability (Henderson 2016).

Regardless, Japan retained some weaknesses as a halal destination, the most crucial of which was the lack of understanding of Islam due to its thin interaction with Muslims (Adidaya 2016). For a long time, Japan had been a monocultural society, with a tiny Muslim minority who had been almost completely ignored. Furthermore, the living cost in Japan was relatively expensive. The raw materials for halal foods were scarce (Yusof & Shutto 2014). Even products declared halal did not always provide a detailed description of the ingredients. The cost for halal certification was high, so was the investment in halal infrastructure. Staff had limited communication skills, largely using Japanese (Henderson 2016).

Thailand is another non-Muslim Asian country that had attempted to take advantage of the economic opportunity offered by halal tourism. It had taken a number of responsive steps to accommodate the needs of Muslim travelers, including the launching of a multilingual Muslim-friendly application, available for Android and IOS, to help Muslim visitors find halal restaurants and hotels, mosques and halal tours, as well as providing travel news and a guidebook (Battour 2018). Thailand established the Al-Meroz Hotel with fully halal facilities and services (Mohsin et al. 2017), followed by four other halal hotels that were launched afterward. More halal hotels could be established in the future (Chandra 2014), especially considering the increased visit of Muslim travelers to the country (Razalli et al. 2012), and the fact that the increase of halal hotels had, in fact, attracted more Muslim tourists (Samori & Rahman 2015).

Thailand was considered one of favorite destinations for Muslims (Rashid et al. 2019), and had welcome visitors from the Islamic world for many decades (Nurdiansyah 2018). The satisfaction level of Muslim travelers who had visited Thailand was relatively high, especially in terms of the availability of halal foods (Sriprasert et al. 2014). Other factors cited to have contributed to the satisfaction of Muslim travelers included security, safety and the destination's socio-cultural attractions. Thailand's tourism strengths were that its people were known to be friendly, its exotic beaches, and its medical services (Wong et al. 2014). Abundant access to halal food had influenced the Muslim travelers to stay longer, intend to return or recommend the destination to others (Mannaa 2019). Thailand had undertaken "gastrodiplomacy" by promoting Thai food to attract more foreign travelers, including Muslims (Muangasame & Park 2019). But Thailand had to face the problem of a shortage of skilled staff (Waehama et al. 2018), especially Muslim staff for halal hotels (Oktadiana et al. 2016). Finding such needed staff was a challenge given the small size of Thailand's Muslim population, and it had to find them overseas which was rather costly (Waehama et al. 2018).

## 2 METHODS

This research surveyed as many published studies on halal tourism as possible as it developed in the four Asian countries. The focus was to identify how the six faith-based needs of Muslim travelers as defined by COMCEC 2016, which were halal food, *Salaah* (Prayer), water usage friendly toilets, Ramadhan services and facilities, facilities with no non-halal activities and recreational facilities with privacy, were satisfied in halal destinations in each country. SWOT analysis was employed to determine the areas of strengths, weakness, opportunities, and threats that each country possesses with regard to its halal destinations, underlining key factors that supported its popularity or the lack thereof.

## 3 RESULTS AND DISCUSSION

The strength of Indonesia's halal destinations lay in that they all catered to most Muslim tourists' needs with an easy access, and the country had received a huge arrival of Muslim tourists with leisure motivation. Inconsistent halal certification, the presence of non-halal products and activities, the low quality of infrastructure to destinations, unskilled staff, and the relatively low arrival of tourists with business and medical/healthcare motivation posed as weaknesses. Opportunities were open in terms of providing improved Ramadan services, locally diverse and unique local experiences and attractions with privacy. The greatest threat was that Indonesia faced fierce competition from Malaysia, which had many similarities but was able to offer better professional services and a more comprehensive halal certification.

While Malaysia was strong in providing easily accessible solutions to tourists' needs and in terms of a high arrival of tourists with all motivations, it had some weaknesses with regard to the presence of non-halal products and activities in the destinations, in addition to the fact that tourism remained largely concentrated in big cities with little privacy. Opportunities abounded in terms of developing business and medical/healthcare tourism and more consistent halal certification for hotel services.

The greatest threats came from Indonesia because it gave more access to rural destinations and from Thailand because it provided robust medical/healthcare tourism at a competitive cost.

Thailand had an abundance of halal foods and prayer facilities and a high arrival of medical/healthcare tourists but remained restricted in the urban areas. In general, the halal destinations in the country suffered from scarce and poorly distributed halal facilities and services, especially in rural areas. Thailand, however, had a great opportunity to improve halal hotel services and to push further the initiatives for halal MICE (Uansaard 2018). The greatest threat to Thailand's halal tourism came from three other countries that had a competitive advantage in almost all aspects.

Japan started to provide more adequate halal facilities and services, although mostly in main cities and in a much lower availability than the other three countries. But the opportunity for further improvements was open, as Japan witnessed a steadily increasing number of inbound Muslim tourists because of the increased promotion of the combination of technological advancement and cultural heritage using social media. The greatest threat was the fact that Japan had to compete fiercely with three other countries that had proved to be more ready, more welcoming and much cheaper.

### 3.1 *Religious and cultural affinity*

Cultural and religious affinity proved to be one of the strongest motivating reason s for inbound Muslim tourists to visit a halal destination in a country. This had been the case with Indonesia and Malaysia, which had received the highest arrival of the four Asian countries studied. The level of inbound tourists' arrival was an empirical yardstick for the strength of any destination (Batabyal & Mukherjee 2018). Was the handling of halal tourism in Indonesia and Malaysia perfect? The answer was "not really." There were some obvious weaknesses at any level, but the perception that the two countries were Muslim countries, or at least Muslim-friendly countries, had been taken for granted, and therefore allowed the ignoring of perceived weaknesses. Halal foods, prayer facilities, and toilets with water were everywhere, from the airports to destinations. Finding a restaurant to eat at or a hotel to stay at that offered no alcohol was not necessarily challenging. Modesty of dress code was more common in public places, as this had not been encouraged or prohibited by Islam. Foreign Muslim tourists felt at home when visiting Indonesia and Malaysia, precisely due the fact that they shared the same religious beliefs, which not only conferred a sense of familiarity and even brotherhood, but also of security and safety. These feelings of familiarity, security and safety were among the most important factors foreigners sought when they were in a new country. In short, the inbound Muslim tourists visiting Indonesia and Malaysia were almost worry-free. The worst that they might find may simply be considered "inconveniences," which could actually enrich the experience being in a foreign land.

The situation would be slightly or even largely different if Muslim tourists visited a non-Muslim or minority Muslim country, such as Japan or Thailand. The host society had different customs and traditions dictated partly by their religious beliefs. In japan, for instance, there was a cultural element, called *Nomikai,* which refers to the possible limitation with which foreigners were unable to socialize as freely with the local population and get involved in their activities (Adidaya 2016). To some extent, a sense of unfamiliarity was accentuated, not only on the part of the visiting tourists but also on the receiving hosts. The two sides could face some awkward situations of how to behave with one-another. Foods were unfamiliar, and raised the question whether they were religiously edible. Prayer facilities were hard to find, toilets with water were similarly rare, as the host society did not have to use water, but papers or tissues, for cleaning. To add to this, the potential prejudices that might have circulated between people from different religious or ethnic backgrounds, such as Islamophobia, which also often came up in the news. In a nutshell, visiting non-Muslim countries, with sets of social and cultural differences, Muslim tourists faced not only some challenges that could make them constantly conscious about everything, including those deficiencies in services that they might readily ignore or forgive in Muslim countries, but also some risks that potentially jeopardize their expectation of gaining meaningful experience and joy during their stay. Thus, religious and cultural affinity was one of the most important factors that had not only shaped the decision to visit in the first place but also the subsequent experience of Muslims when undertaking halal tourism

## 3.2 *Affordability*

Another reason for the high arrival of inbound Muslim tourists was affordability. Of the four countries being studied, Malaysia and Indonesia shared first place among the OIC countries, while Thailand and Japan were ranked second and third among non-OIC countries (GMTI 2019). To some extent, the level of visits was closely correlated with the destinations' relative affordability. Of the four, Indonesia was the most affordable and Japan was the costliest, with Malaysia and Japan being between the two ends of the spectrum. Furthermore, affordability might have compensated for some weaknesses found in destinations, as long they provided the basic halal needs of Muslim travelers, such as halal foods. Economically, the Muslim tourists would be happy to receive more with less money, and they could get just this by visiting Indonesia and Malaysia.

The same affordability thesis also applied in Thailand with regard to medical/healthcare tourism. Thailand had been known for its leading medical treatment, not only in Asia but also globally. Apart from the fact that it had many internationally accredited hospitals and medical personnel with reputable licenses, Thailand's leadership in medical/healthcare tourism had been credited to the fact that it charged much less for the same treatment than if it was done in the West, for instance. Likewise, after the devaluation of the Yen, Japan had witnessed an increase of inbound tourists and had demonstrated a steady rise ever since

## 3.3 *Visa relaxation*

As ordinary as it might appear, the visa relaxation for inbound travelers to enter a country was of significance. Not only did it reduce the financial burden that the prospective tourists could have initially take on, but it also removed the inconvenience that they should go through with the administrative process. In fact, visa freedom policies had increased human mobility (Freier & Holloway 2018), and it was a part of regional alliances and the formation of regional blocs (Czaika et al. 2018). Thus, the entry visa not being required in a number of countries was an expression of alliances that they forged in order to be mutually advantaged and reciprocally benefited from the booming halal industry.

The fact that these four countries had implemented a visa-free entry for their citizens to visit was a sign that they were in an equal position to benefit each other. This policy had been productive in motivating people, including Muslims, to travel across borders. Halal tourism had also been impacted by this visa freedom as more inbound tourist arrival was continually increasing in the four countries. With this ease, prospective tourists needed only to pay for a return flight ticket, and to prepare some amount to spend during their stay in the destination country, without bothering to make an arrangement for appointments with and visit to a foreign consulate or embassy

## 4  CONCLUSION

The results of a SWOT analysis of halal destinations in the four Asian countries suggested three variables as key factors that had shaped tourists' decision to visit and their subsequent experience in destinations, namely religious and cultural affinity, affordability, and visa freedom. The relative correlation that each destination had with the three variables determined its priority by the prospective tourists to be a chosen destination. The more conspicuous the relation that a halal destination had with the three variables, the more likely that it received more visits by inbound Muslim tourists. In this respect, Indonesia and Malaysia had been the most popular halal destinations. Furthermore, it is also suggested, religious and cultural affinity and affordability had compensated for the perceived weaknesses found in destinations, as long as they provided basic halal needs.

# REFERENCES

Adidaya, Y. A., 2016. *Halal in Japan: History, Issues and Problems*. Oslo: Department of Culture Studies and Oriental Languages, University of Oslo.

Batabyal, D. & Mukherjee, S. K., 2018. Determinants of Tourist Visits in Destination Development: An Empirical Analysis of Sikkim. In: *Tourism Marketing: A Strategic Approach*. Oakville, CA & Waretown, NJ: Apple Academic Press Inc, pp. 11–24.

Battour, M., 2018. Muslim Travel Behavior in Halal Tourism. *The Creative Commons Attribution License*.

Chandra, G. R., 2014. Halal Tourism: A New Gold Mine for Tourism. *International Journal of Business Management & Research*, 4(6), pp. 45–62.

Czaika, M., De Hass, H. & Villares-Varela, M., 2018. The Global Evolution of Travel Visa Regimes. *Population and Development Review*, 44(3), pp. 589–622.

Fischer, J., 2011. *The halal frontier: Muslim consumers in a globalised market*. New York, NY: Palgrave Macmillan.

Freier, L. F. & Holloway, K., 2018. The Impact of Tourist Visas on Intercontinental South-South Migration: Ecuador's Policy of "Open Doors" as a Quasi-Experiment. *International Migration Review*, pp. 1–38.

GMTI, 2019. *Global Muslim Travel Index*, s.l., Mastercard-CrescentRating.

Hamid, S. H. A., Aziz, Y. A., Rahman, A. A. & Ali, M. H., 2018. Implementing Big Scale Halal Tourism in Malaysia. *International Journal of Academic Research in Business and Social Sciences*, 8(16), p. 304–318.

Hariani, D., 2017. *Halal Japanese Culinary as Attraction for Muslim Travellers to Visit Japan*. s.l., Advances in Economics, Business and Management Research, pp. 174–176.

Henderson, J. C., 2016. Muslim travellers, tourism industry responses and the case of Japan. *Tourism Recreation Research*.

Jaelani, A., 2017. Industri Wisata Halal di Indonesia: Potensi dan prospek. *Munich Personal RePec Archive, No. 76237*.

Mannaa, M. T., 2019. Halal Food in The Tourist Destination and Its Importance for Muslim Travellers. *Current Issues in Tourism*.

Mohsin, A., Ramli, N. & Alkhulayfi, B. A., 2017. Halal tourism: Emerging opportunities. *Tourism Management Perspectives*, Volume 19, pp. 137–143.

Muangasame, K. & Park, E., 2019. Food Tourism, Policy and Sustainability: Behind the Popularity of Thai Food. In: *Food Tourism in Asia: Perspectives on Asian Tourism*. Singapore: Springer.

Mubarok, F. K. & Imam, M. K., 2020. Halal Industry in Indonesia: Challenges and Opportunities. *Journal of Digital Marketing and Halal Industry*, 2(1), pp. 55–64.

Muhamad, N. S., Sulaiman, S., Adham, K. A. & Said, M. F., 2019. Halal Tourism: Literature Synthesis and Direction for Future Research. *Pertanika J. Soc. Sci. & Hum*, 27(1), pp. 729–745.

Newsroom, T., 2020. *https://www.tatnews.org*

Newsroom, T., 2020. *TAT Governor urges industry patience and solidarity to overcome the COVID-19 crisis*.

Nurdiansyah, A., 2018. *Halal Certification and Its Impact on Tourism in Southeast Asia: A Case Study Halal Tourism in Thailand*. s.l., KnE Social Sciences, pp. 26–34.

OIC, 2017. *Strategic Roadmap for Development of Islamic Tourism in OIC Member Countries*, s.l., The Statistical, Economic and Social Research and Training Centre for Islamic Countries (SESRIC).

Oktadiana, H., Pearce, P. L. & Chon, K., 2016. Muslim Travellers' Needs: What Don't We Know?. *Tourism Management Perspectives*, Volume 20, pp. 124–130.

Osman, N., 2019. *Halal holidays: Sun, sea and seclusion appeal to women*, s.l., s.n.

Pietro, B. P. & Secinaro, S., 2019. The Halal Tourism: A Business Model Opportunity. In: *Islamic Tourism: Management of Travel Destinations*. Oxfordshire: CABI, pp. 192–200.

Rashid, N. R. N. A., Akbar, Y. A. A., Laidin, J. & Muhamad, W. S. A. W., 2019. Factors Influencing Muslim Tourists Satisfaction Travelling to Non-Muslim Countries. In: *Contemporary Management and Science Issues in the Halal Industry*. Singapore: Springer, pp. 139–150.

Razalli, M. R., Abdullah, S. & Yusoff, R. Z., 2012. Is Halal Certification Process "Green"?. *The Asian Journal of Technology Management*, 5(1), pp. 33–41.

Samori, Z. & Rahman, F. A., 2015. Establishing shariah compliant hotels in Malaysia: Identifying Opportunities, Exploring Challenges. *West East Journal of Social Sciences*, 2(2), pp. 95–108.

Samori, Z., Salleh, N. Z. M. & Khalid, M. M., 2015. Current trends on Halal tourism: Cases on selected Asian countries. *Tourism Management Perspectives*.

Sriprasert, P., Chainin, O. & Rahman, H. A., 2014. Understanding Behavior and Needs of Halal Tourism in Andaman Gulf of Thailand: A Case of Asian Muslim. *Journal of Advanced Management Science*, 2(3), pp. 216–219.

Uansaard, S., 2018. Creating the awareness of halal MICE tourism business in Chiang Mai. *Int. J. Tourism Policy,* 8(3).

Waehama, W., Alam, M., Hayeemad, M. & Waehama, W., 2018. Challenges and Prospects of the Halal Hotel Industry in Muslim-majority and Muslim-minority Countries: The Case of Malaysia and Thailand. *Journal of Halal Industry and Services,* 1(1).

Widagdyo, K. G., 2015. Analisis Pasar Pariwisata Halal Indonesia. *The Journal of Tauhidinomics,* 1(1), pp. 73–80.

Wong, K. M., Velasamy, P. & Arshad, T. N. T., 2014. *Medical Tourism Destination SWOT Analysis: A Case Study of Malaysia, Thailand, Singapore and India.* s.l., EDP Sciences.

Yusof, S. M. & Shutto, N., 2014. *The Development of Halal Food Market in Japan: An Exploratory Study.* Kuala Lumpur, Procedia — Social and Behavioral Sciences, pp. 253–261.

*Halal Development: Trends, Opportunities and Challenges – Pratikto et al (eds)*
© *2021 The Author(s), ISBN 978-1-032-03830-8*

# Asking a chatbot for food ingredients halal status

C. Suardi, D. Anggeani, A.P. Wibawa*, N. Murtadlo & I.A.E. Zaeni
*Universitas Negeri Malang, Malang, Indonesia*

N.A.M. Jabari
*Palestine Technical University Kadoorie, Tulkarm, Palestine*

ABSTRACT: For Muslims who live side by side with non-Muslims, understanding halal and haram is significant. This paper is intended to discuss the use of a chatbot in the halal-haram status of food ingredients. The results of this study are a review of foods in various categories including healthy, natural foods and even foods that are not included for consumers. The respondents used consisted of 30 and gave a total assessment of 53.84% of the chatbot that has been built and implemented. The higher the rating given, the higher the quality of the chatbot to identify halal and haram foodstuffs.

*Keywords*: chatbot, food ingredients, halal

## 1 INTRODUCTION

Religion is an important domain to study as it influences everyone (Eid 2013; Mokhlis 2009; Zamani-Farahani & Henderson 2010). There are several religions in this world, one of which is Islam. Islam has its regulations that are supposed to help its faithful to adhere to the right path. The Muslim community globally is vast and will continue to grow. The population is expected to increase to 31.1%, equivalent to 3 billion Muslims in 2060 (Yousaf & Xiucheng 2018).

There is a strict Islamic dietary law. This law specifies the foods should be eaten (halal) or not eaten (haram) (Sin & Sin 2019). Islamic law must follow Al-Qur'an and Sunnah (Abdullah & Ahmad 2010; Suardi et al. 2019). Knowing halal and haram is essential for Muslims who live side by side with non-Muslims. Halal is a term used in Islam to label something permissible, which cannot be excluded for Muslims (Eid & El-Gohary 2015; El-Gohary 2016), and haram is the opposite of halal. It is helpful for Muslims to fully understand the halal or haram animals list because it protects them from carrying out sinful activities. Animals are categorized as haram because they are viewed as impure, unclean, or as the cause of human disease. Examples of animals that are categorized as haram for consumption are cats, dogs, and pigs (Kamarulzaman et al. 2016). They are known as dangerous and do not help human life if they are consumed. The use of technology is not new in the halal discussion. An RFID validates halal food status (Norman et al. 2009; Yang & Bao 2011). Another applicable platform is camera phone barcode scanning for halal dietary detection (Junaini & Abdullah 2008). These technologies indicate halal-status based on food ingredients, yet do not show the law. People may consult a trusted source and goto reliable Islamic scholars to discuss this matter. Finding reliable information sources could be problematic in minority Muslim countries (Bonne & Verbeke 2008). Thus, another technology-driven solution should be delivered to address the problem, and one of the potential solutions is a chatbot. A chatbot is a program that simulates human dialogue with users through chat text messages (Husein et al. 2020; Ranoliya et al. 2017). Its essential role is to help users find answers to their questions. A chatbot operates using rules and

---

*Corresponding author

DOI 10.1201/9781003189282-3

is often considered an artificial intelligence. This service may be used for a variety of things, from practical to enjoyable.

This paper aims to discover the use of a chatbot in the halal-haram status of food ingredients, which is developed using the Extreme Programming (XP) approach. This XP method is easy-to-use, fast and straightforward, especially for developing a small application like a chatbot (Sudarsono et al. 2020). The paper consists of several sections, such as explaining the research method, findings, and discussion.

## 2 METHODS

Co Artificial Intelligence Markup Language (AIML) was used to build the chatbot. It is an XML-based markup language used to build interfaces while keeping the application simple in the program, easy to understand, and maintainable (Satu et al. 2017; Shawar & Atwell 2002). The content of AIML is a collection of questions and responses that can be useful for chatbots to find answers to each user sentence. AIML is used to receive input and search for answers to AIML documents (Wijaya et al. 2020). We used Extreme Programming (XP) to develop the system (Abrahamsson & Koskela 2004). XP is an approach or model for software development that seeks to simplify different development stages to become more versatile and adaptive (Assassa & Dossari n.d.). XP not only focuses on scripts, but also encompasses the whole software development environment (Paulk 2001; Rumpe & Schröder 2014) (Table 1).

### 2.1 User Experience Questionnaire (UEQ)

A User Experience Questionnaire (UEQ) is an easy and efficient tool for measuring User Experience (UX) (Laugwitz et al. 2015; Schrepp et al. 2017). UEQ contains six assessment scales: Attractiveness, clarity, efficiency, dependability, stimulation, and novelty. UEQ has twenty-six question components and seven answer options. A list of questions with a choice of answers or UEQ questionnaires is displayed in Figure 1. Then the data collected will be processed to analyze the data and convert the raw data into information that has useful knowledge value using Data Analysis Tools.

## 3 RESULTS AND DISCUSSION

The system produced in this study is a chatbot application that has a chat appearance. The page will directly interact with the user. The process that will occur is that the user will send a message to the server. The message that has been sent will be matched with an existing pattern in the database. Designing and building this system begins with the design stage, which is to identify the problems described in the previous chapter regarding the importance of reliable information regarding the halal or prohibited nature of an animal for consumption. In the system requirements analysis phase, a literature study on chatbots is carried out and collects research-related data such as lists of permitted and forbidden animals along with propositions that confirm the status of these animals.

The design stage begins with identifying the previous chapter's problem regarding the importance of reliable information regarding the halal or haram of an animal for consumption. In the system requirements analysis phase, a literature study on chatbots is carried out and collects research-related data such as lists of permitted and forbidden animals along with propositions that confirm the status of these animals. The design stage carries out the interface design process of the designed chatbot. The next step of coding is implementing the Artificial Intelligence Markup Language (AIML) method (Wijaya et al. 2020) and inputting data into the built system as diagram of creating a chatbot application shown in Figure 1. The last is testing, where the system is evaluated using methods and questionnaires distributed to obtain assessments from various users. The UI and response testing of the chatbot is shown in Figure 2. The figure illustrates the chatbot response the user's greeting, and in Table 2 shows chatbot interface testing.

Table 1. List of food ingredients halal status.

| Food Ingredients | Status | Hadith or Al-Quran |
|---|---|---|
| Pig | Haram | *Al-Qur'an Surat Al-Maidah* Verse 3 |
| Shrimp | Halal | *Al-Quran Surah Al-Maidah* Verse 96 |
| Bat | Haram | In *Al Mughni* (11: 66) and *Al-Quran Surah Al-A'raf* Verse 157 |
| Snake | Haram | The Prophet *sallallaahu' alaihi wa sallam* said: "Every wild animal with unclean fangs is eaten" (Narrated by Muslim: 1933). |
| Fish | Halal | Fish and Fish Carcasses of HALAL said the Prophet *sallallaahu alaihi wassalam*: "Two carcasses sanctified uTwo carcasses and blood sanctified us. The two carcasses were fish and grasshoppers. While the two bloods are the liver and spleen." (HR. Ibnu Majah) |
| Chicken | Halal | *Al-Quran Surah Al-An'am* Verse 142 |
| Eagle | Haram | The Prophet Muhammad *sallallaahu 'alaihi wasallam* prohibited "the eating of every wild animal and bird with sharp claws" (Narrated by Muslim). |
| Mouse | Haram | The prohibition of animals is because the Prophet *sallallaahu 'alaihi wasallam* ordered us to kill them. "Five fasiqs (nuisance) animals that should be killed both in the halal (haram) and haram lands, namely: snakes, scorpions, ravens, ravens, eagles" (HR. Muslim) |
| Blood | Haram | *Al-Qur'an Surah Al-An'am* Verse 145 |
| Alcohol | Haram | The Prophet Muhammad *sallallaahu 'alaihi wasallam* say "Everything that is intoxicating is alcohol and all alcohol is haram." (HR. Muslim) |
| Horse | Halal | *Al-Quran Surat An-Nahl* Verse 8 |
| Camel | Halal | Al-Quran Surat Al-Maidah Verse 1 "Do you who believe, fulfill the *aqad-aqad*. Livestock for you are permitted, except for those that will be read to you. (That is) by not legalizing hunting while you are doing *Hajj*. Indeed, Allah determines the laws according to His will.) |
| Rabbit | Halal | From Anas *radhiyallahu' anhu*, he said, "We were busy catching rabbits in the *Marru Azh-Zhohran* valley, people tried to catch them until they were exhausted. Then I caught it and I brought it to face Abu Tholhah. So, he slaughtered it and then sent the flesh of his quads or hamstrings to the Prophet *sallallaahu' alaihi wasallam*. Then he accepted it." |
| Lizard | Haram | From Abdur Rahman bin Syibl said: Rasulullah prohibits eating *dhab* (a type of lizard). [Hasan. HR Abu Daud, Al-Fasawi in Al-Ma'rifah wa Tarikh, Baihaqi and written by Al-Hafidz Ibn Hajar in Fathul Bari and approved by Al-Albani in As-Sahihah] |

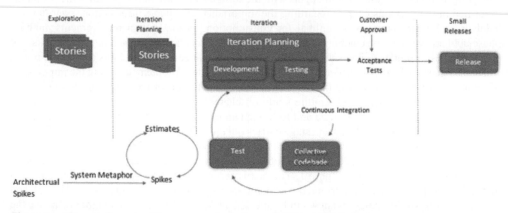

Figure 1. Shows a diagram of the process of creating a chatbot application.

| | 1 | 2 | 3 | 4 | 5 | 6 | 7 | | |
|---|---|---|---|---|---|---|---|---|---|
| annoying | O | O | O | O | O | O | O | enjoyable | 1 |
| not understandable | O | O | O | O | O | O | O | understandable | 2 |
| creative | O | O | O | O | O | O | O | dull | 3 |
| easy to learn | O | O | O | O | O | O | O | difficult to learn | 4 |
| valuable | O | O | O | O | O | O | O | inferior | 5 |
| boring | O | O | O | O | O | O | O | exciting | 6 |
| not interesting | O | O | O | O | O | O | O | interesting | 7 |
| unpredictable | O | O | O | O | O | O | O | predictable | 8 |
| fast | O | O | O | O | O | O | O | slow | 9 |
| inventive | O | O | O | O | O | O | O | conventional | 10 |
| obstructive | O | O | O | O | O | O | O | supportive | 11 |
| good | O | O | O | O | O | O | O | bad | 12 |
| complicated | O | O | O | O | O | O | O | easy | 13 |
| unlikable | O | O | O | O | O | O | O | pleasing | 14 |
| usual | O | O | O | O | O | O | O | leading edge | 15 |
| unpleasant | O | O | O | O | O | O | O | pleasant | 16 |
| secure | O | O | O | O | O | O | O | not secure | 17 |
| motivating | O | O | O | O | O | O | O | demotivating | 18 |
| meets expectations | O | O | O | O | O | O | O | does not meet expectations | 19 |
| inefficient | O | O | O | O | O | O | O | efficient | 20 |
| clear | O | O | O | O | O | O | O | confusing | 21 |
| impractical | O | O | O | O | O | O | O | practical | 22 |
| organized | O | O | O | O | O | O | O | cluttered | 23 |
| attractive | O | O | O | O | O | O | O | unattractive | 24 |
| friendly | O | O | O | O | O | O | O | unfriendly | 25 |
| conservative | O | O | O | O | O | O | O | innovative | 26 |

Figure 2. Shows a list of questions with a choice of answers using UEQ web (modified from Hinderks et al. 2018).

Table 2. ChatBot interface testing.

| No | Testing | Results expected |
|---|---|---|
| 1 | Input Greetings | Can reply to greetings and explain the bot chat system |
| 2 | Input Animal name | Can Reply to the Halal Status of Animal |
| 3 | Input the word Thank You | Can Reply and explain the application |
| 4 | Incompatible Word Input | Can Reply and explain the chatbot system |

Tests were carried out to see the suitability of the system with the expected results. The user enters for several animal names into the system and sees the output that the chatbot replies (Table 2 and Figure 3).

The design stage begins with identifying the previous chapter's problem regarding the importance of reliable information regarding the halal or haram of an animal for consumption. In the system requirements analysis phase, a literature study on chatbots is carried out and collects research-related data such as lists of permitted and forbidden animals along with propositions that confirm the status of these animals. The design stage carries out the interface design process of the designed chatbot. The next step of coding is implementing the Artificial Intelligence Markup Language (AIML) method (Wijaya et al. 2020) and inputting data into the built system as diagram of creating a chatbot application shown in Figure 1. The last is testing, where the system is evaluated using methods and questionnaires distributed to obtain assessments from various users. The UI of the

17

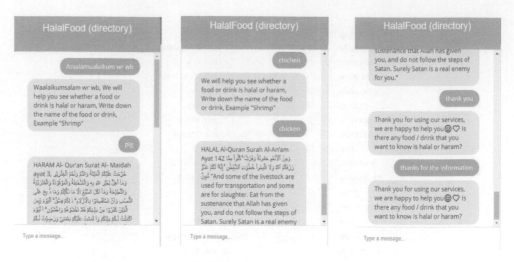

Figure 3.    UI and response testing of the chatbot.

Table 3.    Recapitulation of user questionnaire.

| No | Questions | Very Bad (%) | Not Good (%) | Neutral (%) | Good (%) | Very Good (%) |
|---|---|---|---|---|---|---|
| 1 | Easy to use | 6.7 | 0 | 16.6 | 30 | 46.7 |
| 2 | Creative | 0 | 6.7 | 20 | 23.3 | 50 |
| 3 | Past Response | 0 | 2.7 | 13.8 | 31 | 48.5 |
| 4 | Enjoyable | 0 | 10 | 13.3 | 16.7 | 60 |
| 5 | Helpful | 6.7 | 0 | 3.3 | 16.7 | 73.3 |
| 6 | Correct Answer | 6.7 | 3.3 | 7.3 | 31 | 51.7 |
| 7 | Attractive | 6.7 | 3.3 | 13.3 | 30 | 46.7 |
| Average | | 3.8 | 4.3 | 12.5 | 25.5 | 53.8 |

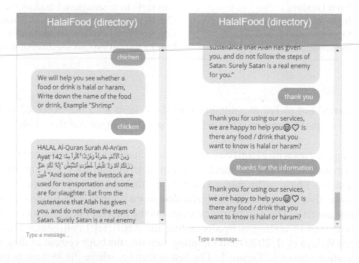

Figure 4.    Chatbot response testing.

chatbot is shown in Figure 3. The figure illustrates the chatbot response the user's greeting, and in Table 2 shows chatbot interface testing.

Table 3 shows that the percentage of user responses using 7 variables, the rating on a 1 or very bad scale, is 3.828571%. The assessment occupies the lowest assessment of the entire evaluation. Furthermore, there was an increase in the ratings of not good, neutral, and right the highest assessment position where users think very good at 53.84286%. In other words, users believe that the chatbot is reliable in terms of explaining the halal-status of food ingredients.

## 4 CONCLUSION

This study concludes that the chatbot system designed and implemented using Artificial Intelligence Markup Language (AIML) and developed with the Extreme Programming (XP) method has been able to help Muslims to provide information on the halal status of an animal for consumption. The expected results can produce overall suitability and provide an outstanding level of satisfaction assessment for 30 respondents. For further research, evaluating user satisfaction and increasing the number of respondents in the current assessment system as well as developing the system according to user needs using the developed methods are encouraged.

## REFERENCES

Abdullah, K., & Ahmad, M. I. 2010. Compliance to Islamic marketing practices among businesses in Malaysia. *Journal of Islamic Marketing*, 1(3), 286–297.

Abrahamsson, P., & Koskela, J. 2004. Extreme programming: A survey of empirical data from a controlled case study. *International Symposium on Empirical Software Engineering – IEEE*, 73–82.

Assassa, G., & Dossari, H. Al. Extreme Programming: A Case Study in Software Engineering Courses. *Strategies*.

Bonne, K., & Verbeke, W. 2008. Muslim consumer trust in halal meat status and control in Belgium. *Meat Science*, 79(1), 113–123.

Eid, R. 2013. Integrating Muslim Customer Perceived Value, Satisfaction, Loyalty and Retention in the Tourism Industry: An empirical study. *International Journal of Tourism Research, Int. J. Tourism Res.*

Eid, R., & El-Gohary, H. 2015. The role of Islamic religiosity on the relationship between perceived value and tourist satisfaction. *Tourism Management*, 46, 477–488.

El-Gohary, H. 2016. Halal tourism, is it really Halal? *Tourism Management Perspectives*, 19, 124–130.

Hinderks, A., Schrepp, M., & Thomaschewski, J. 2018. User Experience Questionnaire.

Husein, M. F. H., Darusalam, U., & Aningsih, A. 2020. Application of the O-Chat Bot Program to Provide Learning Motivation to National University Students Using AIML. *Jurnal Mantik*, 4, 433–440.

Junaini, S. N., & Abdullah, J. 2008. MyMobiHalal 2.0: Malaysian mobile halal product verification using camera phone barcode scanning and MMS. *2008 International Conference on Computer and Communication Engineering*, 528–532.

Kamarulzaman, Y., Veeck, A., Mumuni, A. G., Luqmani, M., & Quraeshi, Z. A. 2016. Religion, Markets, and Digital Media. *Journal of Macromarketing*, 36(4), 400–411.

Laugwitz, B., Schrepp, M., & Held, T. 2015. Konstruktion eines Fragebogens zur Messung der User Experience von Softwareprodukten. *Mensch Und Computer 2006*, (Nielsen 1994), 125–134.

Mokhlis, S. 2009. Relevancy and Measurement of Religiosity in Consumer Behavior Research. *International Business Research*, 2(3).

Norman, A. A., Md. Nasir, M. H. N., Fauzi, S. S. M., & Azmi, M. 2009. Consumer acceptance of RFID-enabled services in validating Halal status. *2009 9th International Symposium on Communications and Information Technology*, 911–915.

Paulk, M. C. 2001. Extreme Programming from a CMM Perspective. *IEEE Software*, 18(6), 19–26.

Ranoliya, B. R., Raghuwanshi, N., & Singh, S. 2017. Chatbot for university related FAQs. *2017 International Conference on Advances in Computing, Communications and Informatics, ICACCI 2017*, 2017-Janua, 1525–1530.

Rumpe, B., & Schröder, A. 2014. Quantitative Survey on Extreme Programming Projects. *Third International Conference on Extreme Programming and Flexible Processes in Software Engineerin*.

Satu, S., Akhund, T. M. N. U., & Yousuf, M. A. 2017. Online Shopping Management System with Customer Multi-Language Supported Query handling AIML Chatbot.

Schrepp, M., Hinderks, A., & Thomaschewski, J. 2017. Construction of a Benchmark for the User Experience Questionnaire (UEQ). *International Journal of Interactive Multimedia and Artificial Intelligence*, 4(4), 40.

Shawar, B. A., & Atwell, E. 2002. *A Comparison Between Alice and Elizabeth Chatbot Systems*.

Sin, K. Y., & Sin, M. C. 2019. Distinguished identification of halal and non-halal animal-fat gelatin by using microwave dielectric sensing system. *Cogent Engineering*, 6(1).

Sudarsono, B. G., Lestari, S. P., Bani, A. U., Chandra, J., & Andry, J. F. 2020. Using an extreme programming method for hotel reservation system development. *International Journal of Emerging Trends in Engineering Research*, 8(6), 2223–2228.

Wijaya, Y. S., Rahmaddeni, & Zoromi, F. 2020. Chatbot Designing Information Service for New Student Registration Based on AIML and Machine Learning. *Journal Of Artificial Intelligence And Applications*.

Yang, Y., & Bao, W. 2011. *The Design and Implementation of Halal Beef Wholly Quality Traceability System*.

Yousaf, S., & Xiucheng, F. 2018. Halal culinary and tourism marketing strategies on government websites: A preliminary analysis. *Tourism Management*, 68(February), 423–443.

Zamani-Farahani, H., & Henderson, J. C. 2010. Islamic tourism and managing tourism development in islamic societies: The cases of Iran and Saudi Arabia. *International Journal of Tourism Research*, 12(1), 79–89.

# Urban human health: A study on food and drink phenomena

N. Athiroh*, M.I. Thohari & A. Sa'dullah
*Universitas Islam Malang, Malang, Indonesia*

ABSTRACT: This research explores food and beverage in Tlogomas Malang, Indonesia. We analyzed perception for food and drink with a social approach and examined borax, formalin, and the heavy metal content of food and beverages in terms of scientific policy. In so doing, both qualitative and quantitative methods were employed. Data was analyzed descriptively. Our identification of food products included 25 food stalls. Sixty foods tested contained borax and formalin, and only one type of food was detected positive for borax and formalin. Findings also suggest that 20 food samples (1.67%) contained sauce, *"terasi"*, and *"petis* are not dangerous. The heavy metal test was carried out on all types of food, namely 80 food samples, and 74 of the 80 food samples tested showed positive results for Cu (copper) with a percentage of 92.5%. (copper). Meanwhile, 42 of the 80 food samples showed positive results for Pb (lead) with a percentage of 52.5% and showed negative results for other metals. The respondents in this research did not complain to vendors and managers. Also, the *"Mamin"* stall on campus has not been tested at the laboratory.

*Keywords*: food, drink, heavy metal, urban, human, health

## 1 INTRODUCTION

Based on an annual report of Indonesian religious life in 2020, the Indonesian Muslim population was approximately 209,12 million people. Thus, the percentage of Muslim people reaches 87% of the total population. The quantity of Muslims in Indonesia has made the country a potential market for halal products (Arifin 2016). The Islamic boarding school is the oldest Islamic institution in Indonesia, contributing to the religion and health fields. The Islamic boarding school also has a significant contribution to the use of halal products (Suharto 2018).

The culture in the *Assalafiyah* Islamic boarding school is not conducive to healthy eating habits. There are so many culinary cafes in the Islamic boarding school area that are cheap, easy to get, tasty, and suitable for people. Which makes them seem like an excellent choice. On the contrary, the food is known to cause food poisoning. The food must have been contaminated by microorganisms like *E. colli* or *Salmonella sp*, and contains "Rhodamine B" colouration in the sauce (Tatebe et al. 2014).

Researchers examined three samples of "tempura" based on all the positive samples' chemistry examination results of all the positive models that contained formalin and borax. A food additive was found to contain borax. Borax becomes toxic when it accumulates in the body. It causes vomiting, fatigue, and renal failure (Pongsavee 2009). Thereby, it is essential to consume halal food based on the Al-Qur'an (Al-Maidah: 88).

## 2 METHODS

The research employed both qualitative and quantitative methods. The qualitative approach included planning preparation, asking for consent and recruiting 50 respondents using a purposive sampling

---

*Corresponding author

DOI 10.1201/9781003189282-4

technique. A quantitative approach was used in the laboratory work, consisting of the formalin test, borax test, and colouration test.

## 2.1 *Research variable*

The variables included in this research were borax, formalin, and heavy metal of food and beverage, respondents' perception of food and drinks was measured using a Likert scale survey, with a social economy character for every individual consisting of age, sex, education, occupation, income, social status, respondents' communication for knowledge of food and beverages, social activity impact of food and beverages, and the distance of their home from food and drinks vendors (Baral 2017).

## 2.2 *Data collection*

The research employed both primary and secondary data. Primary data included a structured-interview using a questionnaire, while the secondary data was gathered from document analysis (Baral 2017).

## 2.3 *Data analysis*

Qualitative data analysis used was an analysis of the inferential test. A hypothesis was proposed to examine a strong or weak relationship between two variables. The quantitative data was tabulated and analyzed using the Anova test (Onwuegbuzie et al.2014).

## 3 RESULTS AND DISCUSSION

The findings document 25 culinary cafes around the Islamic boarding school campus area, with kinds of food tested including deflated tofu (*"tahu gembos"*), chips (*"kerupuk"*), noodles, *"sempol"*, sauce, tuna fish (*"tongkol"*), *"lele"* fish, *"cilok"*, powdered spice, *"cimol"*, meatballs, white tofu, suhun, powder, "pindang", *"batagor"*, a condiment made from pounded and fermented shrimp and small fish (*"terasi"*), and a condiment of fermented fish or shrimp (*"petis"*), and *"Bandeng"* fish.

### 3.1 *Test of borax content*

Among the 60 foods that were tested, only one food showed contamination by borax in the percentage of 1.67%. The danger of borax for the body includes decreasing body weight, vomit, slight diarrhea, appearing spot on the skin, hair falling out, convulsion, and anemia (Pongsavee 2009). Tumbel's (2010) research revealed that wet noodles circulating in Makassar City do not contain borax. Turkez et al. (2013) showed no increase in MNHEPs. Moreover, simultaneous treatments with BX significantly modulated the genotoxic effects of AlCl in rats. BX has beneficial influences and can antagonize Al toxicity (Turkez et al. 2013).

These findings are similar to previous research exploring meatballs with borax. The research found no harmful preservatives in the meatball and showed that borax was not found in beef meatballs. Borax had effects on immune cell proliferation (lymphocyte proliferation) and induced sister chromatid exchange in human chromosomes. The toxicity of borax may lead to cellular toxicity and genetic defects in humans (Pongsavee 2009). The most dominant factor relating to the behaviour of borax usage was health education given to all vendors. The stall vendors' OR value was 2.433 (CI: 90% 1.108-5.342), which means that vendors who never receive health education tend to use borax 2.43 times as much compared to those who have received education (Mujianto et al. 2005).

## 3.2 Test of formalin content

Among the 60 foods tested, only one food showed borax contamination, with a content of 1.67%. The consumption limitation of food matter containing formalin according to the International Program on Chemical safety for an adults is 1.5 – 14 mg per day. For food, 0.2 mg may be eaten per day and for beverage, 0.1 mg per liter of water per day. According to the Occupational Safety and Health Administration (OSHA). The Formalin threshold limitation is 1- 0.1 mM generally (Warner 2013). Exposure to formaldehyde may cause leukemia, particularly myeloid leukemia, in humans. However, results from other investigations are mixed, suggesting caution in drawing definitive conclusions (Hauptmann et al. 2003).

## 3.3 Test of colouration content

We tested 20 samples of food such as sauce "terasi" and "petis" for colouration content. Among the 20 samples of food tested. The impact that occurs can be in the forms of eye irritation, digestive tract irritation and red urine. Most of the "terasi" samples (70%) were found to contain Rhodamine B. Thirteen respondents (43.3%) viewed Rhodamine B as having moderate risk. Most of "terasi's" creators (63.3%) did not know about dangerous colour substances. Meanwhile, 63.3% of the respondents claimed that Rhodamin B is food colouring and is used to colour the "terasi". Adding Rhodamine B to "terasi's" colour makes it look more attractive. It means that most of the sellers have a positive attitude. It also implies that the respondents do not agree with using Rhodamine B for terasi.

This research showed that SF at the elementary school in Kulon Progo District contained excessive food additives and hazardous substances. In fact, 4% of the samples contained excessive sodium benzoate, sorbic acid, and 8% of the samples contained excessive sodium cyclamate's excessive artificial sweetener. SF with boric acid was 3% samples ("cilok", "sosis", "kerupuk rambak") and formalin was 1% samples ("burjo", "cimol"). There was a significant correlation between the SF vendor's education level with knowledge about formalin, boric acid, and artificial sweetener.

## 3.4 Test of heavy metal content

The test of heavy metal by all kinds of food included 80 food samples taken from 25 cafes. In this regard, 74 out of 80 food tested showed positive results to Cu (Copper) with 92.5%, while 42 of 80 food samples showed positive for Pb (lead) with 52% and indicated negative results for other metals. Many children suffer from heavy metal poisoning and this can cause nutrient deficiencies. Consuming heavy metal accumulation with a concentration degree of 100 mg/dL in child blood tends to cause severe disease, comas, convulsions, or death. Previous research revealed that heavy lead in the fish body has served as a contamination indicator.

Recently, many foods are circulated which are unsafe to consume. It is because of the metal contamination such as lead (Pb), mercury (Hg), arsenic (As), and cadmium (Cd). If those metals come into the human body through food, which is consumed by humans, it can cause nerve system problems, brain damage, paralysis, growth resistance, kidney damage, bone fragility, and DNA damage or cancer.

The primary source of lead (Pb) pollution comes from vehicle emissions. Lead contaminations are at various environmental components such as air, water, the ground, and the food itself. Lead can enter the body of a human being through inhalation, ingestion, and through the skin. The different symptoms include infertility in men, trouble menstruating, miscarriages, and even death in women. Lead in a child's body tends to result in continuous anemia and intelligence degradation (Tchounwou et al. 2012).

## 3.5 Analysis of perception to food and beverage using a social approach

To collect respondents' perceptions, a set of questionnaires was distributed to lecturers, official employees, students, and other people (Baral 2017). This includes a) the respondents' description,

23

b) the respondents' need for food and beverages, c) society's character, d) respondents' perceptions on culinary input, e) respondents' perceptions on leading product food and drink brands and by-products from food, f) respondents' perceptions on wasted food and beverages, and g) respondents' perceptions on the impact of the culinary business food and beverages' existence.

In this research, we found that the respondents (42.86%) never conveyed complaints to food and beverage vendors. Most of them (40%) never contacted the vendors and the managers to convey their opinions. This happens due to reluctance since the group of food and beverage culinary leaders in halal, *tayyib*, and *mubarak* stalls never hold meetings with people (71.43%).

Basically, the respondents strongly agree that the main products of food and beverage are of 40% safe ingredients; also supported by having hygiene and sanitation of 42.86%. However, they (34.29%) disagreed very much if the by-product is dangerous for the environment. They agreed (31.43%) if the product is environmentally friendly. They (37.14%) also agreed that if the production process paid attention to the halal aspect and agreed very much if the culinary food and beverages' production process does not cause contamination (28.57%).

The respondents (40%) disagreed with newcomers of culinary food and beverages as they created social distractions in the environment. They (42.86%) also disagreed if criminality increases in the environment. They (48.57%) agreed if Islamic boarding schools work to produce culinary food and beverages. They (40%) agreed that a group of culinary food and bev erage producers help poor people. Our findings have portrayed that Islamic symbolisms have significant influences on the decision to buy food products. It is indicated by a significant p-value, which is <0.005. In other words, the influence of Islamic symbols on buying decisions is 27.3%.

## 4 CONCLUSION

This research has attempted to unravel 25 cafes in the Islamic boarding school campus area with some food such as tofu "*tahu gembos*", chips "*krupuk*", noodle, "*sempol*", sauce, tuna fish ("*tongkol*"), "*Lele*" fish, meatball, white tofu, *suhun*, *siomay*, meatball of fish ("*pentol*") "*cilok*", powder spice, "*cimol*", "*pindang*", "*batagor*", "*terasi*", "*petis*", and "*Bandeng*" fish. Among the 60 tested foods, we found one food positivey-contained borax and formalin, with 1.67% each. Besides, our research portrayed 20 food samples, such as sauce, "*terasi*" and "*petis*" that were tested in terms of colouration content and revealed no endangered colouration in these foods. The massive metal test for all kinds of food included 80 food samples taken from 25 cafes, and 74 from 80 food samples tested showed positive to metal Cu (copper) with 92.5%. In comparison, 42 from 80 food samples showed a positive result of metal Pb (lead) with 52.5% and indicated negative results for other metals. In this research, we also found that our respondents never complained and desired to contact the vendors, waiter/waitress, and managers.

## ACKNOWLEDGEMENTS

We wish to thank the Laboratory of Halal Center, University of Islam Malang, and Ministry of Religious Affairs for providing this research grant through *Penelitian Unggulan Integrasi Keilmuan* scheme (PUIK/12/2016).

## REFERENCES

Arifin, Tajul. 2016. "Misunderstanding of the Indonesian Human Rights Activists on the Application of the Death Penalty." *Asy-Syari'ah* 18(1):185–98. doi: 10.15575/as.v18i1.7607.
Baral, Uma Nath. 2017. "'Research Data' in Social Science Methods." *Journal of Political Science* 17:82–104. doi: 10.3126/jps.v17i0.20515.

Hauptmann, Michael, Jay H. Lubin, Patricia A. Stewart, Richard B. Hayes, and Aaron Blair. 2003. "Mortality From Lymphohematopoietic Malignancies Among Workers in Formaldehyde Industries." *JNCI: Journal of the National Cancer Institute* 95(21):1615–23. doi: 10.1093/jnci/djg083.

Mujianto, Bagya, Anny Victor Purba, N. Sri Widada, and Retno Martini. 2005. "Faktor-faktor yang Mempengaruhi Penggunaan Boraks pada Bakso di Kecamatan Pondok Gede-bekasi." *Indonesian Bulletin of Health Research* 33(4):20279.

Onwuegbuzie, Anthony J., R. Burke Johnson, and Kathleen Mt Collins. 2014. "Call for Mixed Analysis: A Philosophical Framework for Combining Qualitative and Quantitative Approaches." *International Journal of Multiple Research Approaches*. doi: 10.5172/mra.3.2.114.

Pongsavee, Malinee. 2009. "Effect of Borax on Immune Cell Proliferation and Sister Chromatid Exchange in Human Chromosomes." *Journal of Occupational Medicine and Toxicology* 4(1):27. doi: 10.1186/1745-6673-4-27.

Suharto, Toto. 2018. "Transnational Islamic Education in Indonesia: An Ideological Perspective." *Contemporary Islam* 12(2):101–22. doi: 10.1007/s11562-017-0409-3.

Tatebe, Chiye, Xining Zhong, Takashi Ohtsuki, Hiroki Kubota, Kyoko Sato, and Hiroshi Akiyama. 2014. "A Simple and Rapid Chromatographic Method to Determine Unauthorized Basic Colourants (Rhodamine B, Auramine O, and Pararosaniline) in Processed Foods." *Food Science & Nutrition* 2(5):547–56. doi: 10.1002/fsn3.127.

Tchounwou, Paul B., Clement G. Yedjou, Anita K. Patlolla, and Dwayne J. Sutton. 2012. "Heavy Metals Toxicity and the Environment." *EXS* 101:133–64. doi: 10.1007/978-3-7643-8340-4_6.

Tumbel, Maria. 2010. "Analisis Kandungan Boraks Dalam Mie Basah Yang Beredar Di Kota Makassar." *Chemica: Jurnal Ilmiah Kimia Dan Pendidikan Kimia* 11(1):57–64. doi: 10.35580/chemica.v11i1.389.

Turkez, Hasan, Fatime Gcyikoğlu, and Abdulgani Tatar. 2013. "Borax Counteracts Genotoxicity of Aluminum in Rat Liver." *Toxicology and Industrial Health* 29(9):775–79. doi: 10.1177/0748233712442739.

Warner, Melanie. 2013. *Pandora's Lunchbox: How Processed Food Took Over the American Meal*. Simon and Schuster.

# Does *Ramadhan* affect abnormal return?

I. Ashari & Y. Soesetio*
*Universitas Negeri Malang, Malang, Indonesia*

ABSTRACT: This study aimed to test the stock performance during the month of *Ramadhan* and other *Hijriah* months in 1432-1438 H. A comparative study with a quantitative approach to the selected sample data of 16 issuers who are members of the SRI-KEHATI index were used. This study employed a paired sample t-test. The analysis indicated that there are differences in stock abnormal returns each month, but this does not prove that the *Ramadhan* month effect anomaly always occurs. Interestingly, this study proves that, when *Ramadhan* month arrives, it is suspected that investor behavior tends to avoid risk, since existing beliefs and rumors influence investor behavior in Indonesia.

*Keywords*: abnormal return, *Hijriah*, *Ramadhan*, SRI-Kehati Index

## 1 INTRODUCTION

The efficient market hypothesis (EMH) is an ideal framework and is expected to occur in the capital market. Suppose the market reacts quickly and accurately to any information that occurs and immediately forms a new equilibrium price. In that case, investors will find it difficult to obtain a profit level above normal (abnormal return). Fama (1970) divided the efficient market into three forms: the weak-form efficient market, the semi-strong form efficient market, and the strong form efficient market. One form that is often debated and anomalous is the weak form of an efficient market. Market anomalies are observations that cannot be explained by EMH assumptions and cannot be explained by logic (Oncu et al. 2017).

The various forms of weak market anomalies known so far are the Holiday effect, Day of the week, Monday Effect, Weekend effect, Monthly effect, etc. One of the events of the Monthly effect that causes abnormal returns and is still interesting to study is the effect of the month of *Ramadhan (Ramadhan* effect). The month of *Ramadhan* is one of the holy months for Muslims, which demands that every Muslim be prohibited from eating and drinking, and having husband and wife relations during the day from dawn to dusk. This is thought to make people's daily life patterns, especially Muslims, change during the month of *Ramadhan* and other months. Every devout Muslim will carry out more and more quality religious activities such as increasing sunnah prayers, i'tikaf, reading the Al-Quran, *infaq*, and *shodaqoh* (Tan & Ozlem 2018). However, just before *Idul Fitri*, people will buy more needs than usual (zakat, clothes, food, homecoming costs, etc.). This has led to the high level of public consumption during the month of *Ramadhan*. On the eve of Eid al-Fitr in 2013, household consumption even reached IDR 110 trillion or nearly 10% of the total State Budget (APBN), so that the 2013 changes reached IDR 1,683 trillion (Utomo & Herlambang 2015).

During the month of Ramadhan, the high level of public consumption during the month of Ramadhan is thought to have caused consumption or transactions on the stock market to decline or people to pay less attention to investing in the stock market. Mustafa (2011) examined the effects of the month of *Ramadhan* using the KSE-100 index on the Karachi Stock Exchange with

---

*Corresponding author

DOI 10.1201/9781003189282-5

the research period December 1991 to December 2010. He found that the average return during *Ramadhan* is lower and significant, which means the *Ramadhan* effect is present on the stock exchange Karachi. Another study conducted by Utomo and Herlambang (2015) examined the of Eid holidays on average abnormal returns and found that the significant average abnormal return was in the period before the Eid al-Fitr holiday *(Ramadhan)*, both positive and negative. Munusamy (2018) investigated the Islamic calendar effect on the returns of the Shariah and common indices in India. The data include the daily values of the BSE Sensex, Nifty 50, Nifty Shariah, Nifty Shariah 500, and MSCI over the period from January 4, 2010, to March 31, 2016, and found that the average returns of the stock markets are higher during the *Ramadhan* days than non-*Ramadhan* days over the study period.

Meanwhile, previous research conducted by Seyyed et al. (2005) on the phenomenon of the Muslim month, namely the month of *Ramadhan* on the Saudi Equity Market (the largest equity market in the Middle East and the Islamic world) shows that the average rate of return is not affected by the events of the month of *Ramadhan.* there is even a significant decrease in volatility, which indicates a predictable change in price and results in fewer transactions during the holy month. Then Shah et al. (2014) conducted a study using daily data on the KSE 100 index. From January 1, 2010 to December 31, 2012, the study found that religious factors are not related to financial markets. Markets remain the same during the month of *Ramadhan* similar to the other months.

This study was conducted to examine the effect of the month of *Ramadhan (Ramadhan* effect), which is compared with other months by comparing the CAAR value indicator (Cumulative Average Abnormal Return). This study employed shares incorporated in the SRI-KEHATI index, a collection of 25 leading stocks based on an assessment of the implementation of Sustainable and Responsible Investment (SRI) corporate management procedures. This study was conducted in 1432–1438 H (7 December 2010 – 20 September 2017).

## 2 LITERATURE REVIEW

### 2.1 *The cumulative average abnormal return*

Abnormal return is also a form of market reaction to an announcement that contains information. The abnormal return is the advantage of the return that occurs against the normal return (Hartono 2012). Positive abnormal returns identify that the actual stock return during the event period is greater in value than investors' expectations. Negative abnormal returns identify that the actual return in stocks during the event period is smaller than investors' expectations.

This study used a market-adjusted model or a market-adjusted model, assuming that the best predictor for estimating a security return is the market index at that time (Hartono 2012). The formula used to calculate the cumulative average abnormal return is as follows (1).

$$CAAR_t = \frac{\sum CAR_{it}}{k} \tag{1}$$

where $CAAR_t$ = cumulative average abnormal return of all samples in period t; $CAR_{it}$ = cumulative abnormal return shares me in period t; and k = number of samples.

### 2.2 *Efficient market*

Stock markets offer a platform where shares can be valued efficiently (Toit & Hall 2018). Fama (1970) defines the financial market as efficient if the price offered always reflects the information available. According to EMH, the stock price changes randomly where the price at the future is unpredictable. If there is new information, the information appears randomly so that price changes to the information cannot be predicted. EMH is a trading situation in the capital market that shows the relationship between market prices and market forms. The market price is determined by the capital market mechanism, or in other words formed by how much influence the relevant information considered in making investment decisions (Sunariyah 2011).

## 2.3 Capital market anomalies

Anomaly is a generic term in nature. It applies to any fundamental novelty of fact, new and unexpected phenomenon or a surprise about any theory, model, or hypothesis (Latif et al. 2011). An efficient market reports stock prices that fully reflect available information and maintain the return-risk relationship (Hamid et al. 2010; Latif et al. 2011; Toit et al. 2018). However, suppose there is any pattern in share price movements. In that case, it shows that the market is inefficient, and that market anomaly can be exploited to the advantage of investors (Toit & Hall 2018).

Capital market anomaly means a situation in which stock performance or a group of stocks deviates from the assumptions of efficient market hypotheses. Such movements or events that cannot be explained using efficient market hypotheses are called financial market anomalies (Latif et al. 2011). Anomalies in the capital market are divided into four types: company anomalies, seasonal anomalies, event anomalies, and accounting anomalies.

## 2.4 Ramadhan effect

The average stock return tendency on the day before the holiday (pre-holiday return) is higher. The average stock return on the day after the holiday (post-holiday return) is lower, compared to the normal daily rate return or the Holiday effect. The holiday effect in capital market activities on the Indonesia Stock Exchange (IDX) has several holidays, for example, the religious holiday effect, where Indonesia has the largest Muslim majority in the world, and every year the Muslim community always holds a religious festival, namely with the arrival of the month of *Ramadhan* (*Ramadhan* effect) which is followed by the *Idul Fitri* holiday.

Al-Ississ (2010) stated that there are two reasons when the equity market is affected by culture and religious festivals. First, on days of religious and cultural festivals, followers of certain religions do not enter the stock market. Their absence can have a one-sided effect if they are all absent at the same time while the attitude towards investment decisions or have no effect if they differ from the investment point of view. Second, investor mood is a function of religious events that influence investment decisions or investor overreaction behavior towards current information. This condition is also inseparable from psychological factors because psychologically, investors will react more dramatically to insufficient information.

The investor's perspective tends to be more sensitive to differences in months and specific moments. Thus making people more sensitive emotionally to the impact of external influences. Islamic religious warnings such as *Ramadhan* and *Idul Fitri* affect the behavior of the stock market.

Thus, it will cause economic activity to slow down because at that moment, investors will have the opportunity to get the chance to get an abnormal return compared to other months.

## 2.5 Financial behavior

Financial behavior is the study of the influence of psychology on financial practitioners' behavior and the subsequent effect on markets. Behavioral finance is of interest because it helps explain why and how markets might be inefficient (Sewell 2010). Studying, financial behavior must understand the concepts of psychology, sociology, and finance to know the whole idea of economic action. Financial behavior tries to explain and improve investor patterns, including emotional processes and influences in making decisions. Financial behavior also tries to explain what, why, and how finance and investments are based on the human perspective. For example, financial behavior studies financial markets where many market anomalies occur such as the January Effect, speculative market bubbles, and crashes (for example in 1929 and 1987) (Pratama & Wijaya 2020).

## 3 METHODS

This study was conducted using a comparative model, which is used to compare or describe the differences caused by certain events. The situation during religious events, especially the month of *Ramadhan (Ramadhan* effect) is an event in this study, which observes stock price movements, especially on the SRI-Kehati index, to determine whether there is a difference in the cumulative average abnormal return during *Ramadhan* with months other than *Ramadhan,* namely *Muharram, Safar, Rabiul Awal, Rabiul Akhir, Jumadil Awal, Jumadil Akhir, Rajab, Sya'ban, Syawal, Dzulkaidah*, and *Dzulhijjah*. The period in this study used the *Hijri* year, namely 1432-1438 H (7 December 2010-20 September 2017). The reason for choosing a period is that the longer the research period, the better it is in predicting research results, both in terms of mood and the psychology of investors. The population used is all issuers that are members of the SRI-Kehati Index, namely 25 companies. The sample was selected based on specific criteria or purposive sampling by the research plan, namely being continuously incorporated in the SRI-Kehati index and found as many as 16 companies during the observation period of 1432-1438 H (7 December 2010-20 September 2017), and had data complete.

## 4   RESULTS AND DISCUSSION

This study captures an overview of the abnormal return conditions in each month of *Hijriah* calendar. The average abnormal return in this study illustrates that the movement of the average cumulative abnormal return of stocks has increased and decreased every month or, in other words, fluctuated.

The highest cumulative average abnormal return in *Rabiul Akhir* was 0.0214, which means that in the month of *Rabiul Akhir*, the issuers who are members of the SRI-Kehati index will benefit from their investment returns, such as the closing price of ASII (Astra International) in *Rabiul Akhir*. The end of February 9 2015, to be precise, increased from 7,575 to 7,950 to 12 February 2015 (*Bursa Efek Indonesia*, 2015); on the other hand, the lowest cumulative average abnormal return was in the month of Rajab, which was -0.0060, which means that the SRI-Kehati Index issuers have suffered losses on investment returns. Then, if an outline is drawn, the abnormal return movement of the SRI-Kehati index in the observation period appears to be moving downhill.

Table 1 shows the average CAAR value and the results of the CAAR difference test, which in the month of *Ramadhan* has the second-lowest average CAAR after the month of Rajab. For the different test results presented in Table 1, the CAAR difference test between *Ramadhan* and month *Rabiul Awal* and *Dzulkaidah* produced a significance value of less than 0.05, which means that there is a difference between the CAAR of the month of *Ramadhan* and the months of Rabiul Awal and *Dzulkaidah*. While the t value produces a negative value, the CAAR value in the month of Ramadhan is lower than the months of *Rabiul Awal* and *Dzulkaidah*. Whereas for other months apart from *Rabiul Awal* and *Dzulkaidah* there is no significant difference in CAAR.

Table 1.   Paired sample t-test hypothesis test results.

| Period | Mean Ramadhan | Others | t | Sig (2 Tailed) | Conclusion |
|---|---|---|---|---|---|
| *Ramadhan - Muharram* | -0.0051 | -0.0005 | -0.487 | 0.643 | Not significant |
| *Ramadhan - Safar* | -0.0051 | 0.0075 | -2.073 | 0.084 | Not significant |
| *Ramadhan - Rabiul Awal* | -0.0051 | 0.0148 | -5.671 | 0.001 | Significant |
| *Ramadhan - Rabiul Akhir* | -0.0051 | 0.0214 | -1.858 | 0.112 | Not significant |
| *Ramadhan - Jumadil Awal* | -0.0051 | 0.0108 | -1.778 | 0.126 | Not significant |
| *Ramadhan - Jumadil Akhir* | -0.0051 | -0.0022 | -0.337 | 0.748 | Not significant |
| *Ramadhan - Rajab* | -0.0051 | -0.0060 | 0.087 | 0.933 | Not significant |
| *Ramadhan - Sha'ban* | -0.0051 | -0.0037 | -0.143 | 0.891 | Not significant |
| *Ramadhan - Shawwal* | -0.0051 | 0.0048 | -0.694 | 0.514 | Not significant |
| *Ramadhan - Dzulkaidah* | -0.0051 | 0.0092 | -2.494 | 0.047 | Significant |
| *Ramadhan - Dzulhijjah* | -0.0051 | -0.0004 | -0.831 | 0.438 | Not significant |

In this study, most observations on *Ramadhan* events occurred during July. May until July is the most of the month of dividend distribution. At the *Ramadhan* period, the average return was significantly higher and lower depending on the month the *Ramadhan* occurred (Rokhim 2019). The *Ramadhan* effect phenomenon does not happen in Indonesia, especially on the BEI compared to other countries, allegedly due to cultural differences and different conditions during the month of *Ramadhan* between Indonesia and other countries (Rusmayanti et al. 2016). This study's results support the research made by Mustafa (2011) on the effects of *Ramadhan* on the Karachi Stock Exchange – 100 for the period of December 1991 - December 2010. In his research, it was found that the average return in *Ramadhan* is lower and significant. However, there is a positive and significant average return found in the months of *Shawwal* and *Dzulkaidah*. There are two reasons when the equity market is affected by culture and religious festivals. First, on days of religious and cultural festivals, certain religions' followers do not enter the stock market. The impact of their absence can cause one-sided effects if they all have the same attitude towards investment decisions or can have no impact if opinions differ. Second, investor mood is a function of religious events that influence investment decisions or investor overreaction behavior towards current information (Al-Ississ 2010).

The month of *Rabiul Awal* in the observation period of this research, when converted into *Masehi*, will be in December to March with the January period being the majority period of observation as many as 5 periods. It is suspected that during this period, the January effect was in effect, which was an abnormal return that was higher than other months. At the end of the year, there were many stock sales to reduce taxes (tax-loss selling). Investors sold securities that had suffered a loss before the end of the year and would buy back the securities. The act of selling at the end of December and buying in early January causes a decline in prices at the end of the year and an increase at the beginning of the year (Pradnyaparamita & Rahyuda 2017). With such data collected, it is strongly suspected to be the cause of differences in performance.

This strategy is generally carried out by buying leading stocks that drive the index. It is better to buy stocks that are small in number but have a large portion (influence) of the IHSG than stocks that are large in number but have small influence. This is the aim of enhancing performance is used to rearrange the portfolio in a number of periods. This is followed by profit-taking by selling superior stocks that have risen in price and then exchanging them with second-tier stocks as well as taking advantage of the early moments of the year to become an event for investors to start trading enthusiastically as in the beginning of every activity that always looks high in enthusiasm but cannot be maintained in the next period.

Meanwhile, most of the month of *Dzulkaidah* in this study is September. This period coincides with the third quarter of the company's performance, which means that its version has been running for several quarters of the work period. The remaining period has less impact. Based on this, fund managers and investors can predict the company's performance in the current year and begin to prepare the selection of leading stocks to own and second-tier stocks in anticipation to be selected and used for the window dressing and January effect periods. By taking advantage of information asymmetry between traders, it will make players who have fundamental information, especially private information, which is formed, one of which is with trading experience so far, to get returns. In fact, in *Dzulkaidah* or September, the total average value of trading volume is 0.00232 or 34,407,916 (researchers process the data), so that in the month of *Dzulkaidah* the tendency of investors and traders to get returns (Siaputra 2006). Thus, it can be concluded that there is a difference between the abnormal return before the event which tends to decrease and the abnormal return after, which tends to increase.

## 5 CONCLUSION

It is not statistically proven that there is a significant CAAR anomaly in the month of *Ramadhan* with other months other than the month of *Rabiul Awal* and *Dzulkaidah* in companies listed in the SRI-KEHATI Index. This is presumably because most people or investors do not use the

Islamic calendar as the main reference for determining when to invest. However, it is assumed that during *Ramadhan*, investors prefer to use their funds to meet consumption needs and the effects of inflation. The enactment of the January effect is an anomaly return that is higher than in other months. At the end of the year, many fund managers and investors repair and rearrange portfolios by selling stocks to reduce taxes (tax-loss selling), reducing stock portfolio losses before the end of the year, and at the beginning of the year will buy back securities. Meanwhile, during the month of *Dzulkaidah,* which coincides with the period from September to October, the period for information on the distribution of interim dividends occurs. As a result, unconsciously investors and traders will overestimate the opportunity for an abnormal return to coincide with the month of *Dzulkaidah*. Thus, it should be assumed that Indonesia's investors when making decisions are dominated by psychological aspects, moods, and habits.

## REFERENCES

Al-Ississ, M. 2010. The impact of religious experience on financial markets. *Working Paper*. Harvard Kennedy School of Government.

Bursa Efek Indonesia. 2015. *Ringkasan saham*. 02 December 2020. https://idx.co.id/data-pasar/ringkasan-perdagangan/ringkasan-saham/.

Fama, E. F. 1970. Efficient capital markets: a review of theory and empirical work. *The Journal of Finance* 25(2).

Hartono, J. 2012. *Pasar efisien secara informasi, operasional dan keputusan*. Yogyakarta: BPFE-Yogyakarta.

Latif, M., et al. 2011. Market efficiency, market anomalies, causes, evidences, and some behavioral aspects of market anomalies. *Research Journal of Finance and Accounting* 2(9).

Munusamy, D. 2018. Islamic calendar and stock market behaviour in India. *International Journal of Social Economics* 45(11): 1550–1566.

Mustafa, K. 2011. The islamic calendar effect on karachi stock market. *Global Business Review* 13(3): 562–574.

Oncu, M. A., Unal, A. & Demirel O. 2017. The day of the week effect in borsa istanbul: a garch model analysis. *International Journal of Management Economics and Business* 13(3): 521–534.

Pradnyaparamita, N. M. W., Rahyuda, H. 2017. Pengujian anomali pasar january effect pada perusahaan LQ45 di bursa efek Indonesia. *E-Jurnal Manajemen Unud* 6(7): 3513–3539.

Pratama, A. & Wijaya, C. 2020. The impact of *Ramadhan* effect on abnormal return. *SSRG International Journal of Economics and Management Studies* 7(4): 213–221.

Rokhim, R. & Octaviani, I. 2019. Is there a *Ramadhan* effect on Sharia mutual funds? Evidence from Indonesia and Malaysia. *International Journal of Islamic and Middle Eastern Finance and Management* 13(1): 135–146.

Rusmayanti, A., Yusniar, M. W., Juniar, A. 2016. Pengaruh bulan *Ramadhan* terhadap return pasar saham di bursa efek Indonesia (1425 H-1434 H). *Jurnal Wawasan Manajemen* 4(1).

Seyyed, F. J. Abraham, A. Al-Hajji, M. 2005. Seasonality in stock returns and volatility: the *Ramadhan* effect. *Research in International Business and Finance* 19(3): 374–383.

Sewell, M. (2010). *Behavioural finance*. University of Cambridge.

Siaputra, L., Atmadja, A. S. 2006. Pengaruh pengumuman dividen terhadap perubahan harga saham sebelum dan sesudah ex-dividend date di bursa efek Jakarta (BEJ). *Jurnal Akuntansi dan Keuangan* 8(1).

Shah, S. M., Atiq U. R., Syed N. A. 2014. The *Ramadhan* effect on stock market. *European Academic Research* 1(11): 4712–4720.

Sunariyah. 2011. *Pengantar pengetahuan pasar modal edisi keenam*. Yogyakarta: UPP STIM YKPN.

Tan, O. F. & Ozlem, S. 2018. *Ramadhan* effect on stock markets. *Journal of Research in Business* 3(1): 77–98.

Toit, E. D., Hall, J. H., Pradhan, R. P. 2018. The day-of-the-week effect: South African stock market indices. *African Journal of Economic and Management Studies* 9(2): 197–212.

Utomo, V. J. & Herlambang, L. 2015. Efek hari libur lebaran pada emiten yang terdaftar dalam ISSI periode 2011-2013. *Jurnal Ekonomi Syariah Teori dan Terapan (JESTT)* 2(5).

Halal Development: Trends, Opportunities and Challenges – Pratikto et al (eds)
© 2021 The Author(s), ISBN 978-1-032-03830-8

# Analysis of the impacts and challenges of Covid-19 on green *sukuk* in Indonesia

A.F. Suwanan, A.C. Putro, A. Triyanto, S. Munir* & S. Merlinda
*Universitas Negeri Malang, Malang, Indonesia*

ABSTRACT: The Covid-19 pandemic is the biggest challenge in sustainable development. Sustainable development requires large capital, and this capital is obtained from the financial market, which comes from investment. There are various investment activities in Indonesia. As a country where most of the population is Muslim, in investment, they must pay attention to Islamic principles and Islamic law. There are various innovations in Islamic investment or financing, one of which is the Green *Sukuk*. This research aimed to portray the impacts and challenges Green *Sukuk* created during the Covid-19 pandemic in Indonesia. Secondary data using journals, articles, books, Ministry of Finance data, and news provided by the Ministry of Finance as materials and references were used. This study indicated that there are impacts and challenges faced in issuing new Green *Sukuk* during Covid-19 pandemic.

*Keywords*: green *sukuk*, Covid-19, investment

## 1 INTRODUCTION

Economic development aims to increase economic growth, and in carrying out economic development, it is always good to pay attention to the surrounding environment. Economic development that is exploitative can damage the environment (Mubarok 2018). If economic development always pays attention to the environment, it will create a dignified community condition and also establish a just economy. In economic development, there is government intervention, especially to make various policies. As an initial step in economic development to increase economic growth, one of which is an investment. Currently, the development of investment activities is directed towards the Sharia system.

Indonesia, as a predominantly Muslim country, has excellent potential in Sharia investment. In connection with investment policies for economic development that pay attention to the environment and climate change. The Indonesian government issues a financial instrument that focuses on the environment and climate change, namely the Green *Sukuk*. The results of this green *sukuk* issuance will be used to finance green projects in five sectors, namely resilience to climate change for disaster-prone areas, sustainable transportation, energy and waste management, sustainable agriculture, and renewable energy spread across various ministries/agencies (DJPPR 2018). With the existence of Sharia financial instruments, namely Green Sukuk, it will result in the community being wiser in investing because in investing, they also think about using natural resources in a healthy and not excessive manner but can provide maximum results (Karina 2019). Green *Sukuk* is relevant because the use of proceeds from its issuance must be allocated to development that considers the environment so that government efforts to address the problem of climate change can be realized (Suherman 2019). The existence of an environmental Green *Sukuk* is expected to make other investments emulate the Green *Sukuk*.

---

*Corresponding author

DOI 10.1201/9781003189282-6

The Covid-19 pandemic has an impact on all economic activities. One of the impacts is in Islamic finance, so the Sharia financial industry must move quickly to overcome this problem and adapt and make the right strategy to survive the Covid-19 pandemic (Tahliani 2020). Therefore, the implementation of Green *Sukuk* in Indonesia has big challenges. However, Green *Sukuk* must see the problem of the spread of Covid-19 as a challenge that must be transformed into an opportunity to be better at implementing Green *Sukuk* Indonesia.

With this background, this research was conducted to expose the challenges of implementing Green *Sukuk* in Indonesia in facing the Covid-19 pandemic in Indonesia.

## 2 LITERATURE REVIEW

### 2.1 *Investment*

Developments, whether carried out by the state or companies, require capital. There are many ways that this can be done to obtain finance or capital, one of which is by investing. According to Sukirno (1997), investment can be defined as spending on investment or companies to buy goods and production equipment to increase the ability to produce goods and services available in the economy. Meanwhile, according to Sunariyah (2003), investment is obtaining one or more assets usually has a long term with the hope of getting benefits in the future. Investors generally will get feedback from the capital that has been submitted. There are so many investment instruments in Indonesia, including Islamic Sharia. There are so many alternative Sharia investments in Indonesia, namely:

a. Sharia Deposits
b. Sharia Mutual Funds
c. Retail *Sukuk*
d. Sharia stocks and others.

In general, Islamic investment financial institutions divide into two: the money market and the Islamic capital market. Sharia investment is an investment that is based on Islamic laws. In Indonesia, the Islamic financial market is commonly referred to as the interbank money market based on Islamic Sharia principles. Islamic money market (PUAS). There are two financial markets used in this principle.

a. In the form of an interbank *mudharabah* invests certificate (IMA)
b. Bank Indonesia Wadi'ah Certificate (SWBI)

In Indonesia there is also an Islamic capital market which was marked by the establishment of the Jakarta Islamic Index (JII) in 2000.

### 2.2 *Green sukuk*

*Sukuk* in Arabic terms is a form of trust certificate in Sharia type of investment. *Sukuk* has become very popular in the Islamic economy, *sukuk* has been recognized by the Arab community after an expansion of trade to other countries. Sukuk are identified as assets, rights and benefits, and services of project ownership. According to Islamic Sharia Council (2002), *sukuk* are long-term securities based on Sharia principles issued by the issuer. According to the law on State Sharia Securities (SBSN), *sukuk* are securities issued based on Sharia principles, as evidence of the share of investment in SBSN assets, both in rupiah and foreign currencies. The agency in charge of issuing state *sukuk* is a legal entity that has been established and is based on law.

The issuance of sukuk is a new innovation in Islamic finance, and therefore, it is prohibited to use usury and the prohibition of buying and selling as stated in Al-Qur'an Q.S Ali Imran: 130 and Al Baqarah 278-279. Meanwhile, Green *Sukuk* is a Sharia bond issued where all of the proceeds are intended to finance green projects that contribute to climate change mitigation and adaptation and to conserve biodiversity so that it is well preserved. Green *Sukuk* is relevant because in particular, the use of the proceeds from its issuance must be allocated to development that considers the environment so that government efforts to overcome the problem of climate change can be realized (Suherman 2019). The existence of an environmenronmental Green *Sukuk* is expected to encourage green investing. The capital resulting from the Green *Sukuk* issuance is entirely for environmental management so that the natural ecosystem is well preserved along with changes in environmental conditions.

## 3 METHODS

In understanding the impacts and challenges of Green *Sukuk* issuance in Indonesia, this research used a qualitative method. To analyze the result, we conducted a secondary data and literature review through journals, articles, books, Ministry of Finance data sets, and news provided by the Ministry of Finance as materials and references. This study also seeks references through theories that are relevant to the cases and problems being studied.

## 4 RESULTS AND DISCUSSION

The beginning of the Covid-19 virus pandemic occurred in early December 2019, which occurred in Wuhan, China. According to WHO (2020), the coronavirus has the advantage of spreading it very quickly because it spreads only in contact with people infected with the virus so that other people can be infected. This virus generally attacks the respiratory system in humans, especially in the lungs, with various symptoms, ranging from low-grade symptoms to severe symptoms. In research conducted by Krishnamoorthy et al. (2020), there are two types of coronaviruses in the world, namely MERS (middle east respiratory syndrome) and SARS (severe acute respiratory syndrome). Coronavirus is a novel disease or virus that has never been identified by the medical world before. Due to the novelty of the virus, handling protocols are still being researched. The virus is now spreading globally (Djalante 2020).

The rapid spread of the coronavirus has had very serious effects on all aspects of life. In fact, human life in this world must adapt to this virus. The most significant impact occurred in the country's economic sector. Whether they are developed countries or countries that are still developing, all countries have experienced economic downturns. The economic down-turns that have occurred throughout the world, especially in Indonesia, are caused by policies taken to cut the spread of the coronavirus. The implementation of policies in the form of lockdowns is applied throughout the world, especially in Indonesia, so that economic activity has been significantly impacted. As a result of the lockdown applied by countries in the export-import, tourism, and other sectors, the production network has been disrupted. In small areas of a country, Large-Scale Social Restrictions (PSBB) are implemented, resulting in decreased consumption and demand for goods and services, which causes an economic downturn to occur.

According to Nasution (2020), Covid-19 pandemic's impact has resulted in low investor sentiment towards the market, which ultimately leads the market towards a negative trend. Strategic fiscal and monetary measures are urgently needed to provide economic stimulus. As the cases of the Covid-19 pandemic develop, the market fluctuates in a negative direction. Not only that, the slow pace of Indonesia's export activities to China also has a significant impact on Indonesia's economy. The Covid-19 pandemic that occurred in Indonesia and worldwide has resulted in many losses in various sectors. But of the many industries that have suffered badly from the pandemic, there is one economic sector that has no impact from the Covid-19 pandemic. The Islamic Capital Market in Indonesia is one that has seen no impact from the pandemic. Prijambodo (2020) mentions that the impact of Covid-19 was felt in all capital market instruments, but Sharia instruments were more immune to pandemics. The Composite Stock Price Index (IHSG) recorded that the lowest point in five years, namely 3,938 and JCI on 19 March 2019, was recorded at 4,910 and Net Sell reached Rp 7.2 billion on the stock market. In Islamic debt instruments, *sukuk* are observed to be more resilient than bonds. Based on data from the Indonesia Composite Index, a comparison of the *sukuk* and conventional markets shows that the yield on the *sukuk* market has decreased by 2.5 percent while bonds have a deeper decline of 4.46 percent.

### 4.1 *Potential and development of Green sukuk in Indonesia*

The Islamic instrument that has recently been of concern to the Indonesian government is the Green *Sukuk* instrument. The Green *sukuk* has enormous potential in traditional markets with the Islamic financial market and can be the momentum to used prove that the Green *sukuk* is a real Islamic finance influence on the market. From the development of conventional system bonds, Green *Sukuk* is superior. According to Brahim (2018), Green *sukuk* has great potential to expand the market and become a bridge between conventional finance and Islamic finance. Among the many potentials that exist in Green *Sukuk*, several potentials are outlined as follows.

a. In 2029, Indonesia projects 419 trillion IDR in the issuance of sovereign *sukuk*.
b. The increase in Green Bonds in 2029 is estimated to increase by 339.38 USD.
c. Investor demand for Green sukuk is positive from year to year
d. In 2018, domestic investors reached 271.179.

Green *Sukuk* is an investment instrument that is Sharia in nature and moves towards environmental empowerment. Many investors are interested in it because it is in the form of Sharia. However, in the development of Green *Sukuk* in Indonesia, many challenges must be faced. According to the Financial Services Authority (OJK), several challenges must be faced by the Green *Sukuk* in Indonesia, which is engaged in maintaining the environmental sector, including the following:

a. The lack of financial institutions in identifying a risk that might impact the environment due to limited capacity of authority.
b. The lack of awareness of financial institutions causes high risk and lack of government roles in handling environmentally friendly projects.
c. There is a mismatch in the payment period because an environmentally friendly project is a long-term project.
d. Lack of information on environmentally friendly projects.
e. Lack of support from banks in supporting environmentally friendly projects.

Apart from the challenges above, *Sukuk* provides an opportunity for the public or Muslim or non-Muslim investors in Indonesia to invest. The proceeds from the state *sukuk* to be used as APBN financing. There has been an increase in the issuance and contribution of state sukuk in the APBN, it can be proven that the accumulated issuance of state *sukuk* in 2018 has reached Rp. 950 trillion with an outstanding per October 25, 2018 of 657 trillion IDR with an average issuance per year of 79 trillion (Ministry of Finance 2019). The potential of the issuance of Green *Sukuk* is enormous and can build the welfare of the people and the economy of the Indonesian nation. Every time the government issues a *sukuk*, it is always eaten up by market demand, which proves this instrument has much potential.

When Covid-19 pandemic occurred in Indonesia, the Green *Sukuk* instrument was not influenced by this pandemic. The impact of this pandemic on the Green *Sukuk* instrument's issuance was only on the issuance process, which was somewhat delayed because the government was focused on finding ways to deal with current conditions and the social restrictions that had occurred. The Indonesian government on 18 June 2020 reissued the global *sukuk* with a nominal value of 2.5 billion USD or in 35 trillion IDR (exchange rate 14 thousand). According to the Directorate General of Risk and Financing Management (DJPPR) of the Ministry of Finance, this global *sukuk* is issued with a five-year tenor of 750 million USD over 10 years amounting to 1 billion USD and a 30-year tenure of 750 million USD under the *Wakaluh* agreement. *Sukuk* received tremendous demand from global investors to oversubscribe 6.7 times compared to the estimated emission target.

## 4.2 Retail sukuk and savings sukuk

The state *sukuk* market with several investors reaching 243,364 people in the domestic sphere in 2018 with a total of 10 series since 2009 occurred in Retail *Sukuk*. A very large increase, it should be noted that at the beginning of the issuance of retail *sukuk*, the number of investors was only 14,295 (DPS 2018). Apart from the potential Retail *Sukuk* issuance, the government issued Savings *Sukuk* in the domestic market. Initial issuance attracted 11,338 investors issued in 2016 and increased in 2017 to 16,477 investors. The Covid-19 pandemic's impact on the issuance of Retail *Sukuk* and Savings is constrained by the social restrictions implemented. Th rough the Ministry of Finance through the Directorate General of Financing and Risk Management (DJPPR), the government officially offers SR-013 retail *sukuk* series from 28 August 2020 to 23 September 2020, from which the government hopes to get funds from the public of 5 trillion IDR. The results of the *sukuk* issuance by the government are used to increase funds. The coupon offered is 6.05% and can be used within a two-month holding period. It is evident that the potential of Green *Sukuk* in the Domestic market is large. However, the government is still focusing on issuing *sukuk* for the international market. This is meant by the issuance of the Green *Sukuk* in the international market so that the government can attract Green Bond investors in the world to invest in Green *Sukuk*, which has clear benefits.

## 5 CONCLUSION

The Covid-19 virus originates from the city of Wuhan, China. Coronavirus or Corona Virus Diseases is a new disease or virus that has never been identified by the medical world before. Thus, Covid-19 virus spreads very quickly throughout the world due to a lack of knowledge about this virus. The rapid spread of the coronavirus has had very serious effects on all aspects of life. The most significant impact occurred in the country's economic sector. Whether they are developed countries or countries that are still developing, all countries have experienced economic downturns. The Islamic Capital Market in Indonesia is one that has had no impact from the pandemic. When Covid-19 occurred in Indonesia, the Green *Sukuk* instrument was not impacted by this pandemic. The impact and challenges of this pandemic to the Green *Sukuk* instrument's issuance were only in the issuance process, which was delayed because the government was focused on finding ways to deal with current conditions and the social restrictions that occurred.

## REFERENCES

Anggraini, Y. 2018. Peran Green Sukuk Dalam Memperkokoh Posisi Indonesia Di Pasar Keuangan Syariah Global. *Journal of Economic Islamics and Business*. Volume 01.

Brahim, C. T. 2018. The Role of green Islamic sukuk to the promotion of sustainable development objectives. *Journal of the New Economy*, 01, 186–207

Djalante, R., Lassa, J., Setiamarga, D., Sudjatma, A., Indrawan, M., Haryanto, B., Mahfud, C., Sinapoy, M. S., Djalante, S., Rafliana, I., Gunawan, L. A., Surtiari, G., & Warsilah, H. 2020. Review and analysis of current responses to COVID-19 in Indonesia: Period of January to March 2020. Progress in Disaster Science, 6, 100091. https://doi.org/10.1016/j.pdisas.2020.100091

Directoral General of Risk and Financing Management (DJPPR). 2018. Green Sukuk, Instrumen Investasi untuk Pembangunan Berkelanjutan

Direktorat Pembiayaan Syariah (DPS). 2018. Hasil Penjualan Sukuk Negara Ritel seri SR-010, Retriviewed from: https://www.djppr.kemenkeu.go.id/page/loadViewer?idViewer= 7791& action=download

Islamic Sharia Council (2002). Fatwa nomor 32/DSN-MUI/IX/2002, Dewan Syariah Nasional Majelis Ulama Indonesia.

Karina, L. A. 2019. Peluang dan tantangan perkembangan green sukuk di Indonesia. In *Proceeding of Conference on Islamic Management, Accounting, and Economics* (Vol. 2, pp. 259–265).

Krishnamoorthy, S., Swain, B., Verma, R.S. et al. *SARS-CoV, MERS-CoV, and 2019-nCoV viruses: an overview of origin, evolution, and genetic variations.* VirusDis. Springer. 2020. https://doi.org/10.1007/s13337-020-00632-9

Ministry of Finance Republik of Indonesia. 2019. Green Sukuk Issuance Allocation and Impact Report

Mubarok, S. 2018. Islam dan Sustainable Development: Studi Kasus Menjaga Lingkungan dan Ekonomi Berkeadilan. *Dauliyah Journal of Islamic and International Affairs*, 3(1), 129–146.

Nasution, Dito. 2020. Dampak pandemic covid-19 terhdap perekonomian Indonesia. Jurnal Benefit 5(2), (212–224).

Prijambodo, B. 2020. Potensi Investasi Keuangan Syariah Di Era Pandemi. webinar series: Komite Nasional Ekonomi dan Keuangan Syariah (KNEKS), 12 juni 2020

Suherman, S. 2019. Identifikasi Potensi Pasar Green Sukuk Republik Indonesia. *HUMAN FALAH: Jurnal Ekonomi dan Bisnis Islam*, 1(6).

Sukirno, S, 1997. Pengantar Teori Mikroekonomi, Edisi Kedua, Cetakan Kesembilan, PT.Raja Grafindo Persada, Jakarta.

Sunariyah. 2003. Pengantar Pengetahuan Pasar Modal. Yogyakarta. (UPP) AMP YKPN.

Tahliani, H. 2020. Tantangan Perbankan Syariah dalam Menghadapi Pandemi Covid-19. *Madani Syari'ah*, 3(2), 92–113.

WHO. 2020. Report of the WHO-China Joint Mission on Coronavirus Disease 2019 (COVID-19). 16–24. February 2020. World Health Organization.

*Halal Development: Trends, Opportunities and Challenges – Pratikto et al (eds)*
*© 2021 The Author(s), ISBN 978-1-032-03830-8*

# The strategic role of the halal study center in supporting halal product assurance policy

H. Pratikto, Y. Agustina*, M. Diantoro & M. Churiyah
*Universitas Negeri Malang, Indonesia*

F.M. Yusof
*University Technology of Malaysia, Malaysia*

ABSTRACT:    This study aims to describe the strategic role of Halal Studies Centers in supporting government policies in guaranteeing halal products. The research setting was 7 Halal Center s, the main informant was the head of the university's Halal Center. The data was obtained through in-depth review. The research findings indicate that the strategic role of the University's Halal Center is to carry out research on guaranteed halal products that are published to the public, and prepare human resources as halal auditors. The duties of the Halal Supervisor include: supervising the Halal Product Process in the company, determining corrective and preventive actions, coordinating halal product process, assisting the Halal Auditor during the inspection provide assistance to business actors in applying for halal certification, providing recommendations to regulators of halal product assurance, and bringing awareness of halal lifestyle for the community.

*Keywords*:    Halal Center, Halal Study, Halal Policy, Guaranteed Halal Products

## 1    INTRODUCTION

The first position as a country with the largest Muslim population in the world is still occupied by Indonesia. The data from 2019 recorded that there were 209.1 million Muslims in Indonesia, or 87.2% of the total population of Indonesia. This number represents 13.1% of all Muslims in the world (Pew Forum on Religion & Public Life 2019). Accordingly, halal products in the form of goods and services are a necessity for the community. This is because Allah SWT affirms in the words of Surah Al-Baqarah verse 168:

> *"O mankind, eat from whatever is on earth [that is] lawful and good and do not follow the footsteps of Satan. Indeed, he is to you a clear enemy."*

In other verse within An-Nahl, verse 114 precisely, it is also affirmed that:

> *"Then* eat of what Allah has provided for you [which is] lawful and good. And be grateful for the favor of Allah, if it is [indeed] Him that you worship."

This is also reinforced by the results of research by Aransyah et al. (2019), which in their research stated that the halal label on a product encompasses a significant effect of Muslim consumers on the decision to buy a product. Aisyah (2016) and Briliana and Mursito (2017), in their research results, also confirm that Indonesians pay great attention to the halal aspects of cosmetic and personal care products to be purchased. In services such as tourism, public awareness of the importance of implementing the halal concept is also increasing, although there are still pros and cons (Jaelani 2017). Through this explanation, the consumption of halal products is a necessity for the Muslim

---

*Corresponding author

DOI 10.1201/9781003189282-7

community in Indonesia. However, the reality is conflicting these days, where there are still Muslim customers who have difficulty identifying whether the products they purchase are halal or vice versa. Tabrany (in Al-Asyhar 2003) emphasized that in Islam, food prohibition is according to the aspect of faith based on applied teachings. Jalil et al. (2018) also stated that healthy and halal foods go hand in hand.

A purified soul is reflected in consuming halal food. The human mind becomes peaceful and tranquil when consuming halal foods (Zannierah Syed Marzuki et al. 2012). On the other hand, there are physical and mental jeopardy contained in foods that are forbidden by the teaching of Islam (Sabry & Vohra 2013). Halal food has four impacts on human behavior, specifically: a) maintaining the balance of the human soul according to its nature; b) fostering wholeheartedness to uphold and maintain Islamic teachings continually; c) keeping the human mouth open to perpetually saying virtuous words and cleansing the human heart from negative qualities; and d) fostering optimism that God is always with the human being and listens to every prayer. This is based on the hadith narrated by Thabrani from Ibn Abas, precisely:

*"You shall righteously eat to make your prayer be granted"*

The above explanation indicates that bad human behavior such as corruption, lying, behaving arbitrarily, and so on are caused by the consumption of products that are not halal. For them, there is no relation between whatever is consumed and the dimension of spirituality. PEW Research (2011) in its report predicts that the world's Muslim population will amount to 26.5 percent of the world's population by 2030. In addition, according to the 2019 GMTI report, Muslim tourists in 2020 will spend up to a nominal value of 200 US dollars (Churiyah et al. 2020). Currently, a number of countries, both with a majority Muslim population such as Indonesia, Turkey, and Malaysia as well as countries with Muslim minorities such as Thailand, South Korea, and Japan, are attempting to attract Middle Eastern tourists (Rasul 2019). Indonesia as a country with the largest Muslim population globally, has the potential to develop halal tourism (Churiyah et al. 2020; Jaelani 2017; Suradin 2018). Based on this opportunity, the Indonesian government encourages the development of halal tourism in various regions. The Malang local government also agreed this through the Malang program's launch to develop a halal tourism destination (Hakim 2019; Rifa'i 2019). To accelerate the realization of this program, cooperation is required between universities, the industrial stakeholders, and the government.

The State University of Malang (Universitas Negeri Malang) as a university guarantees the quality of halal products, as indicated by the establishment of the Halal Center (PH) in LP2M (Research and Community Services Agency) in February 2019. This Halal Center, among others, aims at producing research and community service projects in the field of outstanding halal products and realizing a healthy organization with a transparent and accountable governance system. Along with the competence of the 21$^{st}$ century generation, covering four Cs (Critical Thinking and Problem Solving, Creativity and Innovation, Collaboration, Communication) and to strengthen the management of the Halal Center team, it is necessary to conduct a SWOT (Strengths, Weaknesses, Opportunities, and Threats) analysis to determine development strategies and synergize with parties to develop a guaranteed halal products, including the established Higher Education Halal Studies Center, the Halal Product Guarantee Management Agency (BPJPH), the Indonesian Ulema Council (MUI), the business community and the industrial stakeholders, both on a national and international scale.

## 2 METHODS

Based on the existing phenomena, variables related to the development of a halal study center were identified. The findings of these variables were then resorted to as the basis for developing the Halal Center management model. This qualitative approach attempted to answer the objectives of this study. This research was conducted at the Halal Product Guarantee Management Agency (BPJPH), Indonesian Ulema Council (MUI), National and International Universities that have developed the Center for Halal Studies. National Universities which were selected were ITS Surabaya Halal

Studies Center, IPB Bogor, and ITB Bandung. Meanwhile, the selected international universities are the Technological University of Malaysia under the Ibnu Sina Institute and the University of Chulalongkorn Thailand. Initial research data collection was carried out by in-depth interviews, document archives, participatory observation, and direct observation.

## 3 RESULTS AND DISCUSSION

The general objective of this research is to generate a management model for the halal study center to strengthen the Management of the Halal Center of LP2M State University of Malang. The following are details of the results of this study.

### 3.1 *The role of university to administer halal product guarantee*

The halal certificate is defined as a recognition issued by the Halal Product Guarantee Agency (BPJPH) based on a written halal fatwa issued by the Indonesian Ulema Council (MUI) to recognize the halal status of a product. Meanwhile, Halal Product Process is a series of activities that include the provision of materials, processing, storage, packaging, distribution, sale, and presentation of products carried out to ensure the product's halal status. Fatwa is given by MUI based on the results of examination and/or examining the halal status of a product. This is carried out by the Halal Inspection Agency (LPH). The implementation of tasks by LPH are supported by a Halal Auditor, specifically parties who have the capability to check the halal status of a product and they have a product halal examination/testing laboratory. Within JPH, business actors are supported by Halal Supervisors who are responsible for PPH.

Based on Figure 1, the role of Higher Education is in the position of the establishment and management of the Halal Center (HC) and LPH. Halal Center (HC) is an institution with the ability to carry out assistance, research and other projects. The Halal Center is assigned to assist MSME. The source of fees is based on the Community Service Fund College; Corporate CSR; and on other independent businesses. HC was established by Government Agencies, Islamic Foundations, State Universities (PTN), and Private Universities (PTS) from Islamic foundations. HC is tasked with assisting, fostering, and supervising Halal Product Assurance, entering the data using the relevant officers (Halal Supervisor) to BPJPH, and managing data (Data Bank). In addition, Figure 1 illustrates the substantial process of Halal Center in carrying out the assistance to MSMEs.

Regarding the Implementation of the Halal Product Guarantee, besides establishing a HC, state universities can establish an LPH. This institution is tasked with checking and testing the halal status of a product. Then, BPJPH issued an LPH accreditation certificate based on the determination of conformity assessment by KAN and MUI. The validity period of the LPH accreditation certificate

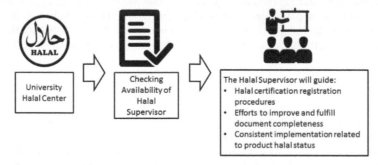

Figure 1.   The role model of University Halal Center.

is five years since it was issued by BPJPH. The existence of Halal Centers in universities plays a role in supporting the halal guarantee policy for the Indonesian government (Akim et al. 2018). In addition, its existence is also an important factor to improve the quality of guaranteed halal products in Indonesia (Gunawan 2020) and an effort to increase the promotion of the halal industry (Yunos et al. 2014). It is expected that the Halal Center in higher education, through its various efforts, can reduce certain public sentiments regarding the concept of halal products (Ainin et al. 2020; Churiyah et al. 2020).

## 3.2  Center for Halal Studies

### 3.2.1  Institut Teknologi Sepuluh Nopember (ITS)
The Center for Halal Studies of ITS, known as ITS Halal Center, was inaugurated in 2016. This Center for Halal Studies is under the ITS Research and Community Services Institution (LP2M). In carrying out its duties, ITS Halal Center always carries out various activities, including education, research, analysis, provision of services to the community, and participating in providing input to the government regarding halal products and regulations. The need for information needed by the public regarding detailed and accurate halal products is the basis for forming the ITS Halal Center institution. Conducting research to find substitutes for haram products commonly used as ingredients for medicines and food is one of the Halal Center's important rol es so far.

### 3.2.2  IPB Halal Science Center (HSC IPB)
Through the provisions contained in the IPB Academic Senate Decree No. 79 / SA-IPB / K / 2017, the Center for Halal Science Studies of IPB (HSC IPB) is one of the centers for science studies at IPB and is under the auspices of the Research and Community Services Agency of IPB. HSC has an essential role in supporting halal inspection and certification, including efforts to develop competent human resources in the field of halal science. This is done to provide benefits to Indonesian and international society.

### 3.2.3  ITB Halal Studies Center (PKH-ITB)
The ITB Halal Studies Center (PKH-ITB) was introduced in 2015, and in its implementation, this institution involved a great number of faculty members from diverse scientific disciplines in the faculties and schools as well as the research centers in ITB. The institutions involved include: a) a School of Life Sciences and Technology (SITH); b) a School of Pharmacy (SF); c) a School of Electrical Engineering and Informatics (STEI); d) an Institute for Research and Community Service (LPPM); e) a Policy Development Agency (SAPPK); f) a Faculty of Mathematics and Natural Sciences (FMIPA); g) a School of Business and Management (SBM); h) a School of Architecture, Planning, and Institute for Innovation and Entrepreneurship Development (LPIK); i) a Tourism Planning and Development Center (P-P2Par); and j) a Nanoscience and Nanotechnology Research Center.

### 3.2.4  Halal Science Center (HSC) Chulalongkorn University of Bangkok, Thailand
The population of Muslims in Thailand is 7 percent of the total population of 66 million. The halal certification initiated in 1949 was solely authorized by the Central Islamic Council of Thailand (CICOT) according to the National Islamic Organization Administration Act 1997. Laboratory tests to facilitate halal authentication and certification in Thailand was first introduced by the Chulalongkorn University Halal Science Center (HSC) in 1999 (Mohd Nawawi et al. 2019). In line with the country's strategy to support the halal affair in Islamic religion in 2004 and Thailand Diamond Halal in 2004, in 2015, HSC-CU aims at achieving its primary vision of being a leader in halal science. This institution is supported by human resources from diverse faculty members of more than 70 people, including 35 professional Muslim scientists possessing more than 150 complete and modern laboratory equipment sets, and a halal inspection agency that obtained ISO 17025:2005

certification and has received a world award as a Halal Science Institution. The existence of this institution plays a very important role in the development of halal tourism destinations in Thailand (Chanin et al. 2015), as well as in the development of halal products in general.

### 3.2.5 *University of Technology Malaysia Halal Studies Center*
Malaysia is one of the leading countries of halal producers in the world, both in terms of institutions that play a role in researching halal products (Azhar 2019), halal food, certification and halal tourism (Henderson 2016; Isa et al. 2018), as well as halal products in general (Nik Muhammad et al. 2009). Halal Center of IIS UTM is under the Center of Research for *Fiqh* Science & Technology (CFiRST), which plays an important role in the development and improvement of halal studies in Malaysia on many aspects of halal products. This institution has the vision to be the center of excellence in scientific and technological *fiqh* investigation.

### 3.2.6 *LPPOM-MUI NTB*
LPPOM-MUI is located in Lombok NTB. This institution has the responsibility of being the leader of the national and even international halal center, considering that NTB has been determined as one of the best halal tourism destinations in the world (Firdausi et al. 2017). The LPPOM-MUI ITB aims at carrying out several duties, specifically: a) establishing and developing Halal standards; b) carrying out Certification of Food Products, medicines, and cosmetics circulating and consumed by the public; c) conducting halal education and raise public awareness to always consume halal products; and d) providing comprehensive information about the halal status of the product from various aspects. NTB Province is a pilot province for halal awareness community movement. NTB is appointed as one of Indonesia's sharia tourist destinations, and NTB has been awarded as the world's best halal tourist destination.

## 3.3 *The Readiness of UM to Support Halal Product Guarantee*

To support achieving its goals, PH-UM hires people who have adequate competence in research, halal process assistance, and halal supervision. The potential of UM in supporting the Halal Product Guarantee can be seen from the availability of human resources, buildings and laboratory facilities owned by UM. There are 49 Lecturers in the Biology Department, 47 from the Chemistry Department, 28 in Industrial Engineering, 17 in the Public Health Department. From the Faculty of Economics, the Department of Management comprises 55 lecturers, 49 people majoring in accounting, and 51 lecturers in development economics. Meanwhile, the existence of a UM central laboratory located in the Faculty of Mathematics and Natural Sciences.

## 3.4 *The Management Model of UM Halal Center*

The Halal Product Guarantee System Management Model, in accordance with the conditions in the State University of Malang, is in the Halal Center under Research and Community Services Agency (LP2M) with the task of conducting research and publishing, mentoring, and educating business actors, and increasing the capabilities of supervisors. Regarding the establishment of the Halal Inspection Agency, the LPH organizational structure consists of the Chancellor (as the supervisor), the Deputy Chancellor (as the steering board), the Chairperson of the Research and Community Services Agency (LP2M) (as the Person in Charge), the Sharia Supervisory Board (as a unit for consultation on halal products according to Islamic sharia), the Head Halal Center (always implementing the LPH activity program), the Central Laboratory Coordinator (as the person in charge of the Laboratory for inspection and product testing in the lab), the Halal Auditor Team (a collection of Halal Auditors as executors of inspection and testing of halal product processes), and Halal Supervisors (as halal product process assistant team).

# 4 CONCLUSION

The State University of Malang (UM) has a high level of readiness in supporting the Halal Product Guarantee program, which is shown by having a Halal Center supported by adequate human resources and fairly complete laboratory facilities. Likewise, UM has made a MoU with BPJPH and is ready to establish a Halal Inspection Agency that supports the straightforward process of certification of halal products for business actors. The UM Halal Center's existence provides benefits to the Indonesian government and, in particular, to the Malang local government in its efforts to launch halal tourism in Malang. Apart from that, the university also obtains benefits to carry out the three pillars of higher education. Business actors, both SMEs and MSMEs, also benefit from providing halal assurance assistance services and the general public to gain easy access to information on halal products. Further research is important to do in order to measure the level of benefit from the existence of the UM Halal Center for the four subjects.

## REFERENCES

Aisyah, M., 2016. Consumer Demand on Halal Cosmetics and Personal Care Products in Indonesia. Al-Iqtishad J. Islam. Econ. 9. https://doi.org/10.15408/aiq.v9i1.1867

Akim, Dr., Dudy Heryadi, R., Utami Dewi, A., Hermawan, C., 2018. Contribution of Higher Education Institutions in Supporting Indonesian Halal Product Guarantee Policy 2019, in: Proceedings of the 2nd International Conference on Social and Political Development (ICOSOP 2017). Presented at the 2nd International Conference on Social and Political Development (ICOSOP 2017), Atlantis Press, Medan, Indonesia.

Al-Asyhar, T., 2003. Bahaya makanan haram bagi kesehatan jasmani dan kesucian rohani. Al-Mawardi Prima.

Aransyah, M.F., Furqoniah, F., Abdullah, A.H., 2019. The Review Study of Halal Products and Its Impact on Non-Muslims Purchase Intention. IKONOMIKA 4, 181–198.

Azhar, A., 2019. The Role of Islamic Religious Institutions in Halal Consumerism in Malaysia: A Review. Int. J. Innov. 7, 14.

Briliana, V., Mursito, N., 2017. Exploring antecedents and consequences of Indonesian Muslim youths' attitude towards halal cosmetic products: A case study in Jakarta. Asia Pac. Manag. Rev., Entrepreneurship and Management in Turbulent Global Environment 22, 176–184.

Chanin, O., Sriprasert, P., Rahman, H.A., Don, M.S., 2015. Guidelines on Halal Tourism Management in the Andaman Sea Coast of Thailand. J. Econ. Bus. Manag. 3, 791–794.

Churiyah, M., Pratikto, H., Filianti, Akbar, M.F., 2020. Halal Tourism: Between Economic Opportunities and Social Acceptance. Nusant. Halal J. 1, 11.

Firdausi, I., Marantika, S., Firdaus, Z.N., Sajidah, R., 2017. Lombok: Halal Tourism as a New Indonesia Tourism Strategy.

Gunawan, A., 2020. Strategy to Improve the Quality of Indonesian Industrial Halal through Smart Investment and Education Systems 4.0. KnE Soc. Sci. https://doi.org/10.18502/kss.v4i7.6857

Hakim, R., 2019. A Review on Halal Tourism: an Analysis on the Parameters. J. Ilm. Ekon. Islam 5, 166–172.

Henderson, J.C., 2016. Halal food, certification and halal tourism: Insights from Malaysia and Singapore. Tour. Manag. Perspect. 19, 160–164.

Isa, S.M., Chin, P.N., Mohammad, N.U., 2018. Muslim tourist perceived value: a study on Malaysia Halal tourism. J. Islam. Mark. 9, 402–420.

Jaelani, A., 2017. Halal Tourism Industry in Indonesia: Potential and Prospects. SSRN Electron. J.

Jalil, N.S.A., Tawde, A.V., Zito, S., Sinclair, M., Fryer, C., Idrus, Z., Phillips, C.J.C., 2018. Attitudes of the public towards halal food and associated animal welfare issues in two countries with predominantly Muslim and non-Muslim populations. PLOS ONE 13, e0204094.

Mohd Nawawi, M.S.A., Abu-Hussin, M.F., Faid, M.S., Pauzi, N., Man, S., Mohd Sabri, N., 2019. The emergence of halal food industry in non-Muslim countries: a case study of Thailand. J. Islam. Mark. 11, 917–931.

Nik Muhammad, N.M., Isa, F.M., Kifli, B.C., 2009. Positioning Malaysia as Halal-Hub: Integration Role of Supply Chain Strategy and Halal Assurance System. Asian Soc. Sci. 5, p44.

Pew Forum on Religion & Public Life, 2019. Religion. Pew Res. Cent. URL https://www.pewre-search.org/projects/religion/ (accessed 10.4.20).

PEW Research, 2011. Global Muslim Population 2030.

Rasul, T., 2019. The trends, opportunities and challenges of halal tourism: a systematic literature review. Tour. Recreat. Res. 44, 434–450.

Rifa'i, M.N., 2019. Integrasi Pariwisata Halal di Kota Malang. Falah J. Ekon. Syariah 4, 194.

Sabry, W.M., Vohra, A., 2013. Role of Islam in the management of Psychiatric disorders. Indian J. Psychiatry 55, S205–S214.

Suradin, M., 2018. Halal Tourism Promotion in Indonesia: An Analysis on Official Destination Websites. J. Indones. Tour. Dev. Stud. 6, 143–158.

Yunos, R.M., Mahmood, C.F.C., Mansor, N.H.A., 2014. Understanding Mechanisms to Promote Halal Industry-The Stakeholders' Views. Procedia – Soc. Behav. Sci. 130, 160–166.

Zannierah Syed Marzuki, S., Hall, C.M., Ballantine, P.W., 2012. Restaurant managers' perspectives on halal certification. J. Islam. Mark. 3, 47–58.

Halal Development: Trends, Opportunities and Challenges – Pratikto et al (eds)
© 2021 The Author(s), ISBN 978-1-032-03830-8

# Production and optimization of protease from *Bacillus Sp.* HTcUM$_{7.1}$ in the process of extracting collagen from milkfish scales as part of the efforts to support the availability of halal collagen in Indonesia

E. Susanti*, Suharti, R.A. Wahab, N.D. Sari & J. Firdaus
*Universitas Negeri Malang, Malang, Indonesia*

ABSTRACT: Protease from *Bacillus sp.* HTcUM$_{7.1}$ was an alternative substitute for pork products and was known to be used in the halal collagen extraction process from milkfish scales. This study aimed to optimize the conditions for the extraction process of the milkfish by a protease from *Bacillus sp.* HTcUM$_{7.1}$. The experimental design in this study used one factorial of completely randomized design (CRD). The research stages were the preparation of milkfish scales, production of crude extract protease, and the optimization of temperature, time, and pH, respectively. The optimum condition was determined based on the highest of soluble collagen protease (PSC) levels measured using the Lowry method. This study's control variable was 5 g of fish scales that had been pretreated with NaOH, crude extract of protease with an activity of 1.56 U with a total volume of 36 mL. The results showed that the enzyme activity resulting from *Bacillus sp.* HTcUM$_{7.1}$ in milk-skim agar were $0.98 \pm 0.48$ U/mL. The protease performance of *Bacillus sp.* HTcUM$_{7.1}$ for the halal collagen extraction process can be increased. A total of 1.56 U crude extract protease from *Bacillus sp. HTcUM$_{7.1}$* in citrate buffer pH 4 with a total volume of 36 mL optimum extract 5 g of milkfish scales at 25°C, for 42 h. The improvement of the protease production method from *Bacillus sp. HTcUM$_{7.1}$* and exploration of its selectivity in various types of fish waste are needed for the development of halal collagen production process methods.

*Keywords*: collagen, protease, milkfish scale, and *Bacillus sp.* HTcUM$_{7.1}$

## 1 INTRODUCTION

Collagen needs in Indonesia rise every year, marked by an increase in the import value of collagen. Indonesia experienced a rise in collagen import value of 2.43% each year between 2012-2016. Collagen importing countries to Indonesia include India, Brazil, the United States of America, Thailand, and China (*International Trade Center* 2017). Collagen from these countries sparks debate amongst academics because the primary source of collagen is from pig and the extraction process includes pepsin from the pig's stomach (Yang & Shu 2014). This calls into question the halal status of the imported collagen. The halal ingredient development and extraction process for collagen is an important issue that is being widely researched.

One alternative for halal collagen ingredients is fish skin, fish bones, and fish scales. Nagai and Suzuki (2000) stated that collagen can be extracted from the skin of Japanese seabass, chub mackerel, bullhead shark, the bones of skipjack tuna, Japanese sea-bass, and yellow sea fish. Another research paper also reported collagen extraction from the skin of the grouper fish skin of parrotfish (Nurhayati et al. 2018) and the scales of tilapia fish (Liu et al. 2016). As the Word of God in Surah Al Maidah verse 26 says: "Sea game and food (which comes) from the sea are lawful. This is also confirmed by the hadith of the Prophet Muhammad SAW, narrated by Ibn Umar

---

*Corresponding author

radhiyallahu 'an-huma. We were allowed two carcasses and blood. The two corpses were fish and grasshoppers. At the same time, the two types of blood are the liver and spleen (Narrated by Ibn Majah no. 3314. Shaykh Al Albani said that this hadith is authentic).

The result of halal collagen extraction is subjected to use of the enzymatic process by altering the enzymatic process by switching from the pepsin from pig stomach to other pro teases, such as proteolytic bacteria. Susanti et al. (2017) has succeed selecting 21 *Bacillus sp,* proteolytic bacteria from *tauco* Surabaya. Based on the microscopic morphology, all isolates shown a rod shape and were presumed to be Bacillus group bacteria. Susanti et al. (2019) stated that from 21 isolates, only *Bacillus sp.* HTcUM$_{7.1}$. had low pathogenicity and produced protease with selective activity similar to milkfish scales. It could be used as a collagen extractor produced with white color, fibers with a size of 20-60 $\mu$m, and collagen type I traits with a denaturation temperature of 145,48°C (Lutfiana et al. 2019).

In general, enzyme activity is determined by its protein structure stability. It is hugely influenced by temperature and pH. Another factor that influences the enzyme is reaction time and relative substrate for the enzyme amount inside the enzymatic reaction mix. Temperature can also increase enzyme activity. The enzyme catalysis reaction rate increases proportionally with temperature. How ever, protein denaturation could happen when the temperature is too high, thereby, the enzyme is losing its activity due to the denaturing of the enzyme molecular structure (Kosim & Putra 2010). Generally, protein experiences denaturation at a temperature above 50°C. The ionization of amino acid sidechains residues at the active center of the enzyme as a catalytic reaction that is influenced by pH. In a specific pH range, an enzyme can perform catalysis optimally. Extremely low or high pH change can result in some changes in ionization residue of the amino acid sidechain at the active center. This condition could lead to changes in enzyme conformation. If the conformation of the structure of the enzyme molecule changes, an enzyme will lose its catalytic ability since it cannot bond with the substrate.

Reaction time gives a unique influence on the enzymatic reaction. In the beginning, enzymatic reaction time runs fast because it has not bonded with the substrate. Then, the reaction rate will slow until constant. Extracted collagen using the pepsin of the skin of a giant croaker *Nibea Japonica* has an optimum time of 8.67 h (Yu et al. 2018), from the skin of yellowfin tuna 23.5 h (Woo et al. 2008), turtle lungs with an optimum time of 15 h (Song et al. 2014) and from the scale of yellowfin tuna 31.02 h (Han et al. 2010). Influence of temperature, pH, reaction time, and substrate concentration in collagen extraction from milkfish scales using protease from *Bacillus sp.* HTcUM$_{7.1}$ has never been researched. Therefore, temperature optimization, pH, and extraction reaction time were conducted in this research.

## 2 METHODS

### 2.1 *Material*

The chemicals used in this research have a degree of p.a. (pro-analysis), except milkfish scales, alcohol, skim milk, tofu liquid waste, destilated water, and methylated spirits.

### 2.2 *Procedure*

The milkfish scales were washed with hot water until they were clean and did not feel slippery. The scales dried at 65°C for 5 h. Fish scales that are not yet to be used are stored in the freezer as stock. *Bacillus sp.* HTcUM$_{7.1}$ was inoculated in skim agar medium with $^3/_4$ quadrant aseptic technique, and incubated at 37°C for 24 h. Then, a total of one single loop of bacteria from *Bacillus sp.* HTcUM$_{7.1}$ was inoculated into 100 mL of production media. Production medium (per L) consisted of 5 g glucose, 5 g MgSO4.7H2O, 5 g KH2PO4, 0.1 g FeSO4.7H2O, and 10% (v/v) tofu liquid waste. The inoculum was shaken with 100 rpm agitation at 37°C for 24 h, centrifuged at 10,000 rpm for 10 min. The supernatants were the crude extract of the protease enzyme, which was then tested for protease activity as referred to in Susanti et al. (2019).

The enzymatic collagen extraction process was modified from Kittiphattanabawon et al. (2010). Milkfish scales were weighed as much as 5 g and soaked in 0.1 M 50 mL NaOH for 6 hours with fast stirring to remove non-collagen protein. NaOH solution was changed every 2 hours. Then the scales were washed with sterile distilled water until pH = 7. The scales were suspended in 6 mL of crude extract of the protease enzyme and 30 mL of 0.5 M acetic acid for 12 hours with continuous stirring. The temperature variations used were 4, 25, and 40°C. The resulting suspension was centrifuged at a speed of 10,000 rpm, 4°C, for 60 minutes. The protein content of the supernatant obtained was measured as Protease Soluble Collagen (PSC). The protein content refers to Lowry's method. Extraction time varied between 6 hours, 12 hours, 18 hours, 24 hours, 30 hours, 36 hours, 42 hours, and 48 hours. The use of acetic acid to suspended scales was replaced with a buffer with various pH values used: 2, 4, 6, and 8.

## 3 RESULTS AND DISCUSSION

### 3.1 *Production of protease crude extract from Bacillus sp. HTcUM$_{7.1}$*

*Bacillus sp.* HTcUM$_{7.1}$ was a proteolytic bacterium isolated from Tauco Surabaya, which can produce selective proteases against milkfish scales and has low pathogenicity. Susanti et al. (2017) reported that the *Bacillus sp.* HTcUM$_{7.1}$ was round and white in shape, and the center was thicker at the edges (Figure 1a). The inoculate of *Bacillus sp.* HTcUM$_{7.1}$ in skim milk agar (SMA) showed the formation of a clear zone around the colony and had the same morphology as pure culture (Figure 1b). The production of crude extract proteases described in Lutfiana et al. (2019) was carried out using one single colony of *Bacillus sp.* HTcUM$_{7.1}$, which was directly inoculated into a production medium containing liquid tofu waste and incubated at 37°C for three days. The results obtained indicated that this method could produce proteases from *Bacillus sp.* HTcUM$_{7.1}$, as the resulting protease activity was 0.98 ± 0.48 U/mL.

### 3.2 *Optimization of the extraction of collagen from milkfish scales enzymatically using a protease from Bacillus sp. HTcUM$_{7.1}$*

Temperature optimization was carried out in two stages of the experiment. In the first experiment, extraction was carried out at temperature variations of 4, 25, 30, and 40°C. The experimental results showed that the PSC at 4 and 25°C was in the form of a clear solution, while at 30 and 40°C it was a thick and cloudy gel. The thick and cloudy gel indicates denaturation of the protein. The natural structure of proteins changes during denaturation so that proteins become insoluble. This condition is not expected in the collagen extraction process. Then optimization is only carried out at temperature variations of 4 and 25°C (Figure 2). At the two temperatures, the PSC level did not differ significantly, although the PSC level at 4°C was lower than 25°C. This result was because at low temperatures, molecules' movement was slower than that induce low product. These results

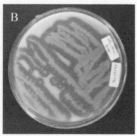

Figure 1. *Bacillus sp.* HTcUM$_{7.1}$ pure culture in NA- Susanti et al. (2019) (A), *Bacillus sp.* HTcUM$_{7.1}$ in SA medium-this research (B).

indicate that the enzymatic collagen extraction process was not taking place in a tight temperature range.

The PSC level increased in proportion to the increase in extraction time between the 6th h and the 42nd h (Figure 3). At the beginning of the reaction, the amount of substrate (S) was excess so that all the enzymes (E) form an enzyme-substrate (ES) complex and immediately convert it into free enzymes (E) and products (P). After the 42nd hour (Figure 3), there was no increase in PSC because the amount of substrate in that condition has started to run out, so there is no increase in the number of products. There was a tendency to decrease the number of products formed. The explanation may be that in these conditions, the buffer could not withstand changes in pH during the catalysis reaction so that it affects the protease conformation, resulting in a decrease in protease activity.

The optimum pH on extraction of collagen protease soluble from milkyfish with protease of *Bacillus sp.* HTcUM$_{7.1}$ was at a pH of 4.0. This result was in line with Song et al. (2014), which stated that the collagen extraction from the turtle's lungs with pepsin was optimum at pH 3.5 and with the papain enzyme at pH 3.0. Based on this, it is possible that the protease in the crude extract of *Bacillus sp.* HTcUM$_{7.1}$ protease has an acidic ionic residue that the optimum protonation for

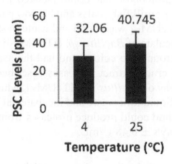

Figure 2.    PSC levels at 4 and 25°C.

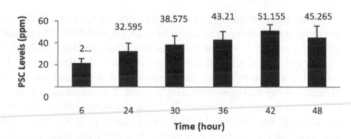

Figure 3.    PSC levels at various extraction times.

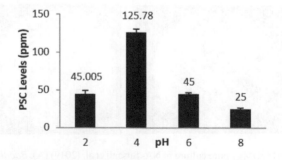

Figure 4.    PSC levels at various pH in the enzymatic extraction process of milky fish scales.

binding with the substrate and the best catalytic reaction is at pH 4.0. Changes in pH below and above pH 4.0 are thought to change the protonation of amino acid residues in the active center. The enzyme conformation can no longer bind to the substrate and cannot carry out its catalytic activity. Protease has an acidic ionic residue with the optimum protonation for binding with the substrate and the best catalytic reaction at pH 4.0. Changes in pH below and above pH 4.0 are thought to change the protonation of amino acid residues in the active center. The enzyme conformation can no longer bind to the substrate and cannot carry out its catalytic activity. However, the results of this study were different from the crude extract of protease *Bacillus sp.* HTR10 isolate, proteolytic bacteria, which came from the Sidoarjo shrimp paste. Research conducted using the same control variable (as much as 1.56 U crude extract of the enzyme in 36 mL buffer with 5 g of fish scales at 25°C) resulted in the highest PSC at a higher pH of pH 6 and extraction time of only 30 hours. It is suspected that the types of proteases produced by *Bacillus sp.* $HTcUM_{7.1}$ and *Bacillus sp.* HTR10 is different, so that they have different characteristics too. *Bacillus sp.* $HTcUM_{7.1}$ was thought to be an acid protease during *Bacillus sp.* HTR10 was a neutral protease (Fitria 2019; Sari 2020).

## 4 CONCLUSIONS

The results showed a total of 1.56 U crude extract protease from *Bacillus sp.* $HTcUM_{7.1}$ in citrate buffer pH 4 with a total volume of 36 mL optimally extract 5 g of milkfish scales at 25°C, for the 42nd h. The improvement of the protease production method from *Bacillus sp.* $HTcUM_{7.1}$ and exploration of its selectivity in various types of fish waste are needed for the development of halal collagen production process methods.

## ACKNOWLEDGEMENT

The authors would like to thank LP2M, State University of Malang, for funding this research through the 2019 PNBP Grant (contract number: 1.3.30/UN32.14/LT/2019).

## REFERENCES

Fitria, N. 2019. Isolat TR-10: *Uji Patogenesis dan Optimasi Suhu serta Waktu Inkubasi Protease dalam Mengisolasi Kolagen dari Sisik Ikan Bandeng.* Unpublish Thesis. Malang: S1 Kimia Universitas Negeri Malang.

Han, Y., Ahn, J., Woo, J., Jung, C., Cho, S., Loo, Y., & Kim, S. 2010. Processing Optimization and Physicochemical Characteristics of Collagen from Scales of Yellowfin Tuna (*Thunnus albacares*). *The Korean Society of Fisheries and Aquatic Science*, 13 (2), 102–111.

International Trade Center. 2017. List of products imported by Indonesia detailed products in the following category: 35 Albuminoidal substances; modified starches; glues; enzymes, 2013-2017, (Online), (https://www.trademap.org.com), diakses 20 September 2020.

Kittiphattanabawon, P., Benjakul, S., & Visessanguan, W. 2010. Isolation and Characterisation of Collagen from The Skin of Brownbanded Bamboo Shark (*Chiloscyllium punctatum*). *Food Chemistry*, 119 (4), 1519–1526.

Kosim, M. & Putra, S. R. 2010. Pengaruh Suhu pada Protease dari *Bacillus subtilis*. *Proceeding Skripsi*. Institut Teknologi Surabaya.

Liu H., & Huang K. 2016. Structural Characteristics of extracted collagen from tilapia (Oreochromis mossambicus) bone: effect of ethylenediaminetetraacetic acid solution and hydrochloric acid treatment. *International Journal of Food Properties*. 19: 63–75.

Lutfiana, N., Suharti., & Susanti, E. 2019. Characterization of Protease Soluble Collagen (PSC) From Milkfish Scales (*Chanos chanos*). *J. Pure App. Chem. Res.*, 2019, 8 (3), 245–254.

Nagai, T., & Suzuki, N. 2000. Isolation of collagen from fish waste material-skin, bone, and fins. *Food Chemistry*, (68): 277–281.

Nurhayati., Tazwir., & Murniyati. 2013. Ekstraksi Dan Karakterisasi Kolagen Larut Asam Dari Kulit Ikan Nila (*Oreochromis Niloticus*). *JPB Kelautan dan Perikanan*, 8(1): 85–92.

Sari, A. Y. 2020. *Optimasi Konsentrasi Substrat, Pengaruh Ion Na$^+$ dan K$^+$ pada Ekstraksi Kolagen secara Enzimatis Menggunakan Ekstrak Kasar Protese Isolat TR-10.* UnpublishThesis. Malang: S1 Kimia Universitas Negeri Malang.

Susanti, E., Lutfiana, N., Suharti., & Retnosari, R. 2019. Screening of proteolytic bacteria from tauco Surabaya based on pathogenicity and selectivity of its protease on milky fish (*Chanos chanos*) scales for healthy and halal collagen production. *IOP Conference Series: Materials Science and Engineering.*

Susanti, E., Ramadhan, R. H., & Fatma, F. 2017. Isolasi dan Seleksi Bakteri Proteolitik Potensial dari Tauco Surabaya. *Prosiding Seminar Nasoinal Kimia dan Pembelajarannya 2017*, 172–182.

Song, W., Chen, W., Yang, Y., Li, C., & Qian, G. 2014. Extraction Optimization and Characterization of Collagen from The Lung of Soft-shelled Turtle *Pelodiscus sinensis. International Journal of Nutrition and Food Science*, 3 (4), 270–278.

Woo, J., Yu, S., Cho, S., Lee, Y., & Kim, S. 2008. Extraction Optimization and Properties of Collagen from Yellowfin Tuna (*Thunnus albacares*) Dorsal Skin. *Food Hydrocolloids*, 22, 879–887.

Yang, H., & Shu, Z. 2014. The Extraction of Collagen Protein from Pigskin. *Journal of Chemical and Pharmaceutical Research*, 6 (2), 683–687.

Yu, F., Zong, C., Jin, S., Zheng, J., Chen, N., Huang, J., Chen, Y., Huang, F., Yang, Z., Tang, Y., & Ding, G. 2018. Optimization of Extraction Conditions and Characterization of Pepsin-Solubilised Collagen from Skin of Giant Croaker (*Nibea japonica*). *Marine Drugs*, 16, 29.

Halal Development: Trends, Opportunities and Challenges – Pratikto et al (eds)
© 2021 The Author(s), ISBN 978-1-032-03830-8

# Legal studies on the halal product certification institution in Indonesia after the enactment of the law on the halal product guarantee

M. Sakti*, Pujiyono & M.N. Imanullah
*Universitas Sebelas Maret, Surakarta, Indonesia*

ABSTRACT:    The implementation of the halal product certification in Indonesia is closely related to efforts to fulfill consumer rights, especially Muslim consumers. It is not only limited to consumer goods but also used goods such as clothing and services of tourism and hotels, which many people expect to be halal-guaranteed. Halal is also related to religious rules, but many people already understand that halal products are healthy and clean. In addition to protecting consumers, it can also increase the economic income of business actors. The halal certification process in Indonesia, which was initially under the authority of the Indonesian Ulema Council, has now shifted its authority after the enactment of the Halal Product Guarantee Law. The Indonesian Ulama Council is no longer the only institution authorized to certify halal products. However, there is a new institution under the Ministry of Religion, namely the Halal Product Guarantee Agency. In 2019, the scope of the Halal Product Guarantee Law should be adequate. Nevertheless, the news institution that organizes halal product certification is experiencing several obstacles, which then creates an impression in the community that the institution is not ready for and is ineffective in carrying out its role in implementing halal food certification in Indonesia.

*Keywords*:    certification, institution, halal, product, guarantee

## 1    INTRODUCTION

Indonesia is a country with the largest Muslim population globally, with more than two hundred and seven million people or around 87% of the total Indonesian population. Despite being a pre-dominantly Muslim country, Pancasila remains the Indonesian nation's ideology without reducing Islamic law values. In the 1945 Constitution Article 28E paragraph (1), it is stated that everyone is free to embrace a religion and worship according to their religion. This is the basis that the state must protect and guarantee to its citizens right to worship and ability to carry out the beliefs taught in their religion.

Islam is one of the religions recognized in Indonesia. Therefore, the state must protect and guarantee the Islamic religion's adherents to carry out worship by the Sharia taught by their religion. One thing that is taught is that Muslims are obliged to consume halal products by Islamic law. The word halal comes from Arabic, which means *to let go* and *not to be bound*. Etymologically, halal means things that can be done because they are free or not bound by the provisions that prohibit them or can mean anything free from harm (Putra 2017). Halal Products are also products that have been declared halal according to Islamic law.

The halal industry has experienced rapid development in recent years. The halal lifestyle that is synonymous with Muslims has spread to various countries, even to countries with minority Muslim populations. Halal is a universal indicator of product quality assurance and living standards (Gillani et al. 2016). However, the development of food technology has been so rapid that the use of ingredients in food processing has varied considerably—the development of the use of ingredients

---

*Corresponding author

DOI 10.1201/9781003189282-9

with specific desirable properties at low prices has become common. The problem is the number of food ingredients, both the primary raw materials and the addictive ingredients, which make it difficult to determine the ingredients' halal origin. The clarity of information on a product consumed is a halal product, or the legal provisions are not clear. Therefore, the existence of a halal label is an important thing to include in every food product. A halal label will make it easier for consumers to identify food products so that even without in-depth knowledge of food additives that allow using haram ingredients, consumers will feel safe when consuming food that is labeled halal. Even so, there are still Muslim consumers who ignore the halal label. Therefore, consumers' perceptions about the importance of information on a product's halal-ness need to be known so that the products released by the food industry are safe products for consumption both in terms of nutritional value and halal-ness.

Along with the law changes that regulate the guarantee of halal products in Indonesia, this cannot be separated from the lack of legal certainty regarding the guarantee of halal products in Indonesia. A few cases have happened to consumers in Indonesia, especially Muslim consumers, who feel disadvantaged because they have to consume foods that are not halal. It can be said that there is no halal label on the products they consume. For example, take meatballs containing pork in several parts of Indonesia and meningitis vaccine cases for pilgrims containing pork enzymes. If we look at this case, it can be said that one of the factors is an inconsistent, overlapping, and non-systemic regulatory system. To provide protection for consumers, the most fundamental thing is that the process of should be an mandatory for every business actor, but the regulations that apply at that time are still voluntary (Hasan 2014).

Of the various regulations that pertain to the guarantee of halal products at that time, it turned out that they still did not fully provide legal certainty to consumers, especially the Muslim community, regarding these regulations. Then in 2006, the People's Representative Council of the Republic of Indonesia (DPR RI), through a proposed initiative, proposed a Bill (RUU) on the Halal Product Guarantee, which was under discussion for eight years. In the end, the bill was passed by the President of the Republic of Indonesia to become law number 33 of 2014 on Halal Products Guarantee, on October 17, 2014. After being ratified and promulgated by law number 33 of 2014, Article 67 paragraph (1) of ingredients,' the law states for products circulating and traded in Indonesia's territory takes effect five years from the enactment of the law. It means that on October 17, 2019, the law came into effect. This confirms that every product that enters circulates and is traded in Indonesia's territory is required to be certified halal. In this case, through the Ministry of Religion of the Republic of Indonesia, the Government is responsible for organizing the Halal Product Guarantee (JPH). Thus, a Halal Product Guarantee Body was formed, which is domiciled under and responsible to the Minister of Religion of the Republic of Indonesia as the organizing Institution for guaranteeing halal products.

A legal concept that contains values, principles, concepts, and basic principles for the formation of the legal substance content of halal products is based on the real needs of the Indonesian people to improve the standard of living, economy, and freedom of practice of religious values and not solely based on business needs. Especially at the expense of the interests of the consumer community. Halal certification in Indonesia's National legal system has a central position because halal certification is contained in law number 33 of 2014, as part of the legal system, namely a legal substance with legal force and legal certainty. That is imperative. Moreover, this is an effort to protect consumers in Islamic Law (Putra 2017).

In connection with the enactment of law number 33 of 2014 on Halal Product Guarantee, this implies a change in the procedure and registration system for halal certification from voluntary to mandatory, which creates a new Institution called the Halal Product Guarantee Body. Thus, there was a shift in the authority for halal certification, which initially belonged to the Indonesian Ulema Council for Food, Drug and Cosmetics Assessment Institute (LPPOM MUI), recently shifted to the Halal Product Guarantee Body, and also affects the process of implementing halal product certification in Indonesia. Based on this background, the writer sees a problem. Namely, are the regulations on the Halal Product Guarantee Law's implementing bodies for halal product certification useful in implementing the halal certification procedure?

## 2 CERTIFICATION OF HALAL PRODUCTS AS CONSUMER PROTECTION

When a country enters the stage of a welfare state, demands for government intervention through the formation of laws that protect the weak are extreme (Fishmen 1986). All halal products must follow Sharia law, including the logistical process. Logistics service providers play an essential role in ensuring that raw materials, raw materials, packaging, storage, and transportation of halal products are carried out properly to not be contaminated with non-halal products (Soon et al. 2017).

Talking about halal product certification, of course, is closely related to consumer protection. As consumers, the Indonesian people need certainty regarding the product's halal-ness, especially food. Consumers have a greater risk than business actors. In other words, consumer rights are very vulnerable. This is due to consumers' weak bargaining position, so that consumer rights are very vulnerable to being violated (Makarim 2003). Oughton and Lowry (1997) view consumer protection law as a modern phenomenon typical of the 20th century, but as confirmed in legislation, legal protection for consumers began a century earlier. The theory of legal protection is a theory related to providing services to the community.

Awareness in registering for halal product certification has not been a major priority for business actors in Indonesia. Compared to other countries, especially countries with most non-Muslim people, many have started to pay attention to and seek their products' halal label. This is not only to fulfill the rights of Muslim consumers, but these business actors also have the view to have more value in trade competition in the Muslim market. When the Muslim community sees that its products are halal certified, they will believe that already halal products are more controlled for quality. They are cleaner and healthier.

Halal certification is a certification process for a product or service by the provisions of Islamic Sharia. Halal certification was first carried out in the United States in the 1960s to guarantee Muslims living in non-Muslim countries to fulfil their needs according to the provisions of their religion. Halal is a mandatory requirement for every product and service consumed by Muslims and is currently considered a product quality standard. Halal quality standards are applied to providing and producing food, cosmetics, medicines, and medical products and is applied to services related to these halal products (Noordin et al. 2014).

Halal can be defined as a quality standard by Islamic Sharia law and is used in every activity carried out by Muslims (Bohari et al. 2013). Muslims choose halal products and services as a form of obedience to Islamic law. Although halal is closely related to Muslims, it does not mean that halal products only come from Muslims. Consumers of halal products from countries with Muslim minority populations have experienced a significant increase in recent years. The quality of halal products is why non-Muslims use halal products (Samori et al. 2016).

In the General Explanation of the Halal Product Guarantee Law, it is explained that the purpose is to ensure that every religious adherent worship and carry out their religious teachings, provide protection and assurance regarding the halal-ness of products consumed and used by the public by the principles of protection, justice, legal certainty, accountability and transparency, effectiveness and efficiency, and professionalism. Besides, implementing a halal product system aims to provide comfort, security, safety, and certainty of halal products' availability for the public to consume and use products and increase added value for business actors to produce and sell Halal products.

Some of the provisions in it, among others, in Article 4 regulates the obligation of halal certification, namely that products that enter, circulate, and are traded in Indonesia must be certified halal. Moreover, Article 5 mandates a new Institution to implement halal certification, namely the Halal Product Guarantee Body, which then requires a Presidential Regulation that will regulate its functions, main tasks, and organizational structure.

## 3 THE IMPLEMENTING INSTITUTION HALAL PRODUCT CERTIFICATION

The Halal Product Guarantee Law regulates each institution's authority that plays a role in the halal certification process of a product. The Halal Product Guarantee Body must cooperate with the Indonesian Ulema Council to carry out halal product certification and supervise Indonesia products. When viewed for law enforcement, the existence of this Halal Product Guarantee Law provides more guarantees of protection and legal certainty than before the existence of this law.

After the implementation of the Halal Product Guarantee Law, Commission VIII of the House of Representatives of the Republic of Indonesia issued a statement that they considered that the Halal Product Guarantee Body under the Ministry of Religion was still slow in carrying out its duties. It was reported that the Head of the Halal Product Guarantee Body, Sukoso, believed that the cause of the slow performance was because the Indonesian Ulema Council seemed to slow down the Halal Auditor certification process, which was a requirement for the establishment of a Halal Inspection Institution. However, this statement was refuted by evidence that the Indonesian Ulema Council had carried out certification for 142 auditors from the Halal Inspection Institution for the Indonesian Ulema Council for Food and Drug Studies (LPPOM MUI) and other candidates for the Halal Inspection Institution from several companies. Other problems also arise from business actors regarding the implementation of halal certification in Indonesia. Many business actors claim to be increasingly confused about the procedure for applying for halal certification because the community has been familiar with the Indonesian Ulema Council as the holder of the halal certification authority. Business actors have also complained about the new halal certification procedure, which is considered more complicated and confusing than the previous procedure.

In the author's opinion, in terms of its traditional strength, the Halal Product Guarantee Law is good enough to accommodate Muslim consumer rights protection. If it is further reviewed regarding the transfer of authority of the implementing Institution for halal product certification, especially to the effectiveness and easy access of business actors in applying for halal certificates, basically the process of implementing halal product certification can run optimally if there is harmonious cooperation between the institutions concerned.

The author compares the implementation of halal product certification with Sharia banking in Indonesia, which involve more than one institution, namely Bank Indonesia, which is under the Ministry of Finance, the Financial Services Authority, and the National Sharia Council of the Indonesian Ulema Council. Islamic banking in Indonesia began to be known with Bank Muamalat Indonesia (BMI) establishment in 1991. The legal basis for the operation of Islamic banking is law number 21 of 2008 on Islamic Banking. Article 1 point 7 states that what is meant by Sharia banks are banks that carry out their business based on Sharia principles. What is meant by Sharia principles by Article 1 point 12 of the Sharia Banking Act is the principle of Islamic law in banking activities based on fatwas issued by institutions that have the authority to determine fatwas in the field of Sharia.

The institution that is authorized to issue fatwas on Sharia financial institutions, including Sharia banking, is the National Sharia Council established by the Indonesian Ulema Council. At the beginning of its formation, it was motivated to realize the aspirations of Muslims regarding economic issues and to encourage the application of Islamic teachings in the economic/financial sector, which was carried out by the guidance of Islamic law, as a measure of efficiency and coordination of scholars in responding to issues related to economic problems/finance. Various problems or cases requiring a fatwa will be accommodated and discussed together to obtain a standard view in their handling by each Sharia Supervisory Board in Islamic financial institution.

Based on its establishment's background, the National Sharia Council of the Indonesian Ulema Council has the authority to issue Islamic economic fatwas, including Islamic banking. The fatwa is used as a guideline for Islamic financial institutions in carrying out their activities. Judging from the current practice of Islamic banking activities in Indonesia, the Islamic banking fatwa from the National Sharia Council of the Indonesian Ulama Council is binding on Sharia banking institutions. This means that Islamic banking institutions must follow the fatwa issued by the National Sharia Council of the Indonesian Ulema Council in carrying out their activities. Currently, nearly 100 fatwas related to Islamic financial institutions, including Islamic banking, were issued by the National Sharia Council of the Indonesian Ulema Council. So, if the Islamic banking institution does not follow or deviate from the fatwa, a warning can be given. Besides, the National Sharia Board can also propose to the Financial Services Authority to take firm action regarding these violations. Thus, the implementing institution for halal product certification that has been mandated in the Halal Product Guarantee Law will be useful in carrying out its duties if there is good cooperation between related institutions, such as cooperation between institutions in the field of Islamic banking was argued by the authors. The law has also emphasized each institution's primary duties and functions so that the relevant institutions need to carry out their respective roles.

## 4 CONCLUSION

The implementation of halal product certification is closely related to the protection of Muslim consumers in Indonesia. As support, an agency for implementing halal product certification was needed, which must be stated in a legal product. In 2014, the Government of Indonesia issued the Halal Product Guarantee Law and was effective in 2019. The Halal Product Guarantee Law mandates the obligation for every product circulating in Indonesia to be halal certified. Also, there is a transfer of authority to the implementing agency for halal certification, initially only carried out by the Indonesian Ulema Council, now under the Ministry of Religion's authority with an institution called the Halal Product Guarantee Agency. However, it turns out that the role of the Indonesian Ulema Council was not wholly transferred and removed. However, the Halal Product Guarantee Agency must coordinate and cooperate in carrying out halal product certification in the process. According to the author, implementing halal product certification can run if there is harmonious cooperation between the institutions concerned. An example is the institutions related to the field of Islamic banking. So, the implementing agency for halal product certification that has been mandated in the Halal Product Guarantee Law will be useful in carrying out its duties if there is good cooperation between related institutions, such as cooperation between institutions in the field of Islamic banking. This is supported by the Halal Product Guarantee Law, which regulates each implementing agency's primary duties, functions, and roles for halal product certification.

## REFERENCES

Apriyantono, Anton, and Nurbowo. 2003. Panduan Belanja dan Konsumsi Halal. Jakarta: Khairul Bayaan.
Asshiddiqie, Jimly. 1994. Gagasan Kedaulatan Rakyat dalam Konstitusi dan Pelaksanaannya di Indonesia. Jakarta: PT. Ichtiar Baru Van Hoeve.
Fishmen, Karen S. 1986. An Overview of Consumer Reporting Service. Maryland: National Law Publishing Corporation.
Fuady, Munir. 2000. Hukum Bisnis dalam Teori dan Praktik. Bandung: Citra Aditya Bakti.
Hartono, Sri Redjeki. 2000. Kapita Selekta Hukum Ekonomi. Bandung: Mandar Maju.
Hasan, KN. Sofyan. 2014. "Kepastian Hukum Sertifikasi dan Labelisasi Halal Produk Pangan." Jurnal Dinamika Hukum 227–238.
Hasan, Sofyan. 2014. Sertifikasi Halal dalam Hukum Positif, Regulasi dan Implementasinya di Indonesia. Yogyakarta: Aswaja Pressindo.
Makarim, Edmon. 2003. Kompilasi Hukum Telematika. Jakarta: PT. Raja Grafindo Persada.
Mujiyono, Slamet. 2016. "Perlindungan Konsumen: Regulasi Bisnis." Jurnal Ekonomi dan Bisnis Islam.
Muslimah, Siti. 2012. "Label Halal pada Produk Pangan Kemasan dalam Perspektif Perlindungan Konsumen Muslim." Jurnal Yustisia 86–95.
Oughton, David, and John Lowry. 1997. Textbook on Consumer Law. London: Blackstore Press Ltd.
Putra, Panji Adam Agus. 2017. "Kedudukan Sertifikasi Halal dalam Sistem Hukum Nasional sebagai Upaya Perlindungan Konsumen dalam Hukum Islam." Jurnal Ekonomi dan Keuangan Syariah 150–165.
Rasyidi, Lili. 1988. Filsafat Hukum. Bandung: Remadja Karya.
Sayekti, Nidya Waras. 2014. "Jaminan Produk Halal dalam Perspektif Kelembagaan." Jurnal Ekonomi dan Kebijakan Publik 193–209.
Shidarta. 2004. Hukum Perlindungan Konsumen Indonesia Edisi Revisi. Jakarta: Grasindo.
Siradj, Asep Syarifuddin Hidayat dan Mustolih. 2015. "Sertifikasi Halal dan Sertifikasi Non Halal pada Produk Pangan Industri." Jurnal Ahkam 199–210.
Syawali, Husni, and Neni Sri Imaniyati. 2000. Hukum Perlindungan Konsumen. Bandung: Mandar Maju.

*Halal Development: Trends, Opportunities and Challenges – Pratikto et al (eds)*
© 2021 The Author(s), ISBN 978-1-032-03830-8

# The influence of halal awareness and halal certificate on purchase intention through brand image

J.D.H. Saputro, I.P. Wilujeng* & H. Pratikto
*Universitas Negeri Malang, Malang, Indonesia*

ABSTRACT: The aim of this study was to determine the direct effect of halal awareness, halal certificates, and brand image on consumer purchase intention. This research used an explanatory design, which distributed questionnaires by using a purposive sampling method and obtained 152 responses from students in Malang, East Java, Indonesia. The findings indicate that halal awareness, halal certificates, and brand image influence consumer purchase intentions. Furthermore, halal awareness has also been shown to have an indirect effect on buying interest through a brand image. The conclusion of this study is beneficial for halal product marketers as a reference to pay more attention to the importance of the existence of halal certificates in marketing products.

*Keywords*: halal awareness, halal certificate, purchase interest, and brand image

## 1 INTRODUCTION

Population growth and increasing salary in Muslim-majority countries reveal a growing demand for halal products globally (Endah 2014). The potential demand in the market is estimated at $2.3 trillion globally (World Halal Forum 2013), and in 2019 is expected to increase to $3.7 trillion. The need for halal products comes from consumers of all religious backgrounds. It motivates several different nations such as Indonesia, Japan, Malaysia, and Thailand, who are competing to become centers of halal products (Global Pathfinder Report 2011).

Muslims are the majority population in Indonesia. It made people aware of halal products. It means consumers have enough halal knowledge in finding and consuming halal products under Islamic law. The awareness of halal has a big influence on the interest in purchasing halal products (Aziz & Chok 2013) because Muslims believe that consuming halal food and products is healthy, a blessing and an obligation for Muslims. The word halal comes from Arabic, which means permissible based on Islamic law (Issa Z 2009; Borzooei & Asgari 2013). Halal items have been claimed halal in accordance with Islamic law, which has conditions for the provision of materials, processing, storage, packaging, distribution, sale, and presentation of halal products (Hidayat & Resticha 2019). The command to eat halal food, drinks and items has been declared in the word of Allah SWT, as in *QS. Al Baqarah: 168, QS. Al Maidah: 88, QS. An Nahl: 114*. In Indonesia, to ensure the halal-ness of a product, there is an agency that exists to certify products as halal. This agency issues a halal certificate as a guarantor of the halal-ness of a product.; as proof that the food or product is free from prohibited elements.

The increase in halal awareness in Indonesia has prompted the government to prepare a law regarding the halal assurance of a product (Sucipto 2009). The law, No. 33 of 2014 about the Guarantee of Halal Products, must also be proactive in strengthening the basis for cooperation and development of halal diplomacy, both at the national and global levels. This cooperation is strengthened, among others, with related ministries and institutions, as well as the Halal Inspection Agency (LPH) and the Indonesian Ulema Council (MUI). The purpose of the MUI Halal

---

*Corresponding author

DOI 10.1201/9781003189282-10

Certification, according to LPPOM MUI, is to provide certainty about the status of halal-ness, to reassure consumers. Companies wishing to obtain an MUI halal certificate (those in the processing industry, catering, slaughterhouses, and restaurants) must register and meet the requirements for halal certification.

Halal awareness is one of the factors that can have an impact on buying interest. In a study conducted by Widyaningrum (2019), halal awareness has a positive and important impact on consumer purchase intentions. The outcome of this research was also proved by Izzuddin (2018). It stated that halal awareness affects purchase intention. Other results proved that every individual who has halal awareness has a significant positive effect and tends to increase his buying interest (Aziz & Chok 2013; Nurcahyo & Hudrasyah 2017; Faturrohman 2019; Fitria et al. 2019). Products that have a halal certificate have guaranteed safety and halal-ness. The existence of this halal certificate and label also makes consumers interested in these products. Aziz and Chok (2013) found that the presence of a halal certificate in a product has a significant positive effect on purchase intention. Other studies also supported the result, and they said that products that have a halal certificate have a significant positive impact and increase consumer purchase intention (Majid et al. 2015; Nurcahyo & Hudrasyah 2017; Nugraha et al. 2017; Sumartik et al. 2019; Widyaningrum 2019; Faturrohman 2019).

## 2 LITERATURE REVIEW

### 2.1 The effect of halal awareness on brand image

Halal awareness is the level of understanding of Muslims in knowing issues connected to the concept of halal. This understanding includes a perception of halal products and how they were prepared under Islamic halal standards (Shaari & Arifin 2010; Ahmad et al. 2013). In addition, halal awareness can be seen as a process of getting information to extend the level of awareness about what Muslims can consume (Ambali & Bakar 2014). Research conducted by Izzuddin (2018) stated that halal awareness and food ingredients influenced purchase intention.
H1. Halal awareness has a significant positive influence on brand image.

### 2.2 The influence of halal certificate on brand image

The guarantee of the halal-ness of a product based on Islamic law is indicated by a halal certificate and a halal logo (Janah 2018). Products that have passed the halal test by MUI are permitted for consumption. Products that have a halal certificate are guaranteed in their safety and halal-ness. The existence of this halal certificate and label also makes consumers interested in these items. Aziz and Chok's (2013) study stated that a product's halal certificate has a positive impact on brand image.
H2. The halal certificate has a significant positive effect on brand image.

### 2.3 The effect of halal awareness on purchase intention

Halal awareness includes knowledge about halal products and how their preparing process is suitable under Islamic halal standards (Shaari & Arifin 2010; Ahmad et al. 2013). Consumers need proper information about halal and haram products like food, beverages, medicines, cosmetics and other various types of consumer goods (Abdul-Talib & Abd-Razak 2013). Research conducted by Izzuddin (2018) and Fitria et al. (2019) revealed that halal awareness influences buying interest.
H3. Halal awareness has a significant positive effect on purchase intention

### 2.4 The effect of halal certificates on purchase intention

The guarantee of the halal-ness of a product based on Islamic law is indicated by a halal certificate and a halal logo (Janah 2018). Research conducted by Nurcahyo and Hudrasyah (2017) stated that there is a relationship between halal certification and purchase intention. The study concludes that a certificate of halal affects purchase intention. Another study done by Aziz and Chok (2013) also showed that the awareness of and certification of halal products was an important factor in increasing the trade of halal products in other religious communities.

H4. There is a significant positive effect of a halal certificate on purchase intention.

## 2.5 *The influence of brand image on purchase intention*

Brand image is an illustration of the overall understanding of a brand, which is shaped by information and experiences relate to it. The brand's image is connected to attitudes in the form of beliefs and references to a brand. Consumers who have a good opinion of a brand were likely to purchase (Setiadi 2007). This theory is supported by some studies (Dianita & Arifin 2018), it stated that the image of the brand influences the items likelihood of purchase. It is also mentioned by Hermanto (2019) that brand image is the most dominant influence variable on consumer purchase intention in Xiaomi Smartphone products.

H5. Brand image has a significant positive effect on purchase intention.

## 2.6 *The impact of halal awareness on purchase intention through brand image*

Halal awareness is the level of consumer understanding to seek and use halal products under Islamic law (Shaari & Arifin 2010). For that reason, halal awareness can be perceived as the process of getting information about items that one can consume (Ambali & Bakar 2014). Research conducted by Izzuddin (2018) stated that halal awareness influences purchase intention.

H6. Halal awareness has a positive and significant effect on buying interest through brand image.

## 2.7 *The effect of halal certificates on purchase intention through brand image*

To ensure the halal-ness of a product in Indonesia LPPOM MUI issues a halal certificate that proves that the food or products are free from prohibited elements. A study by Nurcahyo and Hudrasyah (2017) unveiled that halal awareness influences buying interest in instant noodles in Bandung.

H7. The halal certificate has a significant positive effect on buying interest through brand image.

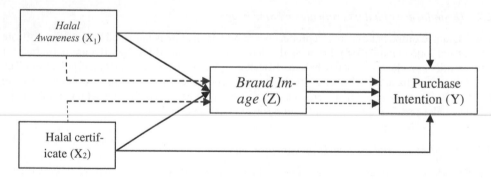

Figure 1. Conceptual framework.

## 3 METHODS

A total of 152 students who have interest in buying toiletry products in Malang City were requested to fill out an online questionnaire, which is adopted and developed from previous studies. It had previously been tested for reliability and validity. All statements were measured on a 5-point Likert scale, ranging from 1 (strongly disagree) to 5 (strongly agree). Path Analysis was used to analyze the data.

# 4 RESULTS

The findings suggest that most consumers who have an interest in buying halal toiletries are aged between 20-22 years, with a total of 105 respondents (69.1%). Meanwhile, based on gender, most of the respondents were 111 women (73%).

Path analysis showed modeled coefficient 1, the beta value of the Halal Awareness (X1) variable on Brand Image (Z) was 0.211 with a significance of 0.010 and the Certification (X2) variable on Brand Image (Z) was 0.427 with a significance of 0.000. Thus, the equation model in the path analysis can be observed, namely: equation model 1, the causal effect between the Halal Awareness (X1) and Halal Certification (X2) variables on Brand Image (Z) can be described by the following equation (1).

$$Z = \beta 1 X 1 + \beta 2 X 2 + e \quad \text{or} \quad Z = 0.211 X 1 + 0.427 X 2 + e \tag{1}$$

Path analysis model summary 1: the R Square value was 0.325. It indicated the influence of Halal

Awareness (X1) and Halal Certification (X2) on Brand Image (Z) is 32.5%, while the remaining 67.5% is a contribution from other variables that cannot be explained because it is not included in the scope of this study.

Referring to the results of the path analysis model coefficient 2, the beta value of the Halal Awareness (X1) variable on Buying Interest (Y) is 0.165 with a significance level of 0.015, the beta value of the Halal Certification variable (X2) on a Purchase Intention (Y) of 0.189 with a significance value of 0.009, and the beta value of the Brand Image (Z) variable on Purchase Intention (Y) of 0.522 with a significance value of 0.000. Path analysis model summary 2 result, the R Square value is 0.553. This shows that the variable Halal Awareness (X1), Halal Certification (X2), and Brand Image (Z) has a contribution of 0.553 or 55.3% to the Purchase Intention (Y) variable, while the remaining 44.7% is the contribution of the variable. Others cannot be explained because they are not included in the variables used in this study. Thus, the equation model in path analysis includes equation model 2, the causal effect between the variables of Halal Awareness (X1), Halal Certification (X2), and Brand Image (Z) on Purchase Intention (Y) described as the following equation (2).

$$Y = \beta 3 X 1 + \beta 4 X 2 + \beta 5 Z + e \quad \text{or} \quad Y = 0,165 X 1 + 0,189 X 2 + 0,522 Z + e \tag{2}$$

Table 1.   Path analysis result.

| No | The influence of variable | Causal influence | |
| | | Direct | Indirect |
| --- | --- | --- | --- |
| 1 | X1 To Z | 0.211 | |
| 2 | X2 To Z | 0.427 | |
| 4 | X1 To Y | 0.165 | |
| 5 | X2 To Y | 0.189 | |
| 7 | Z To Y | 0.522 | |
| 8 | X1 To Y Through Z | | 0.110 |
| 9 | X2 To Y Through Z | | 0.222 |

\* Note the influence of:
1) X1 to Y through Z: $\beta 1 \times \beta 5 = 0.211 \times 0.522 = 0.110$
2) X2 to Y through Z: $\beta 2 \times \beta 5 = 0.427 \times 0.522 = 0.222$

It can be concluded that first, Halal Awareness (X1) and Halal Certification (X2) affect Brand Image (Z). Second, Halal Awareness (X1) and Halal Certification (X2) affect Purchase Intention (Y). Third, Brand Image (Z) affects the Halal Certification (Y). Fourth, Halal Awareness (X1) and Halal Certification (X2) affect Purchase Intention (Y) through Brand Image (Z).

Based on the T-test result the Halal Awareness variable has a t value of 2.625 with a significance value of 0.010 and the variable Halal Certificate has a t value of 5.310 with a significant value of 0.000. it means that the two variables have a significant value smaller than 0.05, which is the standard significance of the t test. This indicated that the two independent variables have a positive influence, it can be concluded that Halal Awareness and Halal Certificate have a significant effect on a positive coefficient on Brand Image. Thus, Halal Awareness (X1) and Halal Certification (X2) has a direct effect on Brand Image (Z). Furthermore, Halal Awareness (X1) and Halal Certification (X2) and Brand Image (Z) on Purchase Intention (Y) count of 2.635 with a significance value of 0.009, and the variable Brand Image (Z) have a value of t count of 7.807 with a significance value of 0.000. This shows that the three independent variables have a positive influence. It can be concluded that halal awareness (X1), Halal Certification (X2), and Brand Image (Z) have a significant effect on the direction of the positive coefficient on Purchase Intention (Y).

Regarding the Sobel test, it can be concluded that the relationship between Halal Awareness (X1) and Purchase Intention (Y) through the Brand Image (Z) showed a two-tailed probability of $0.01287195 < 0.05$. From these results, it can be said that the variable Brand Image (Z) can mediate the relationship between the Halal Awareness (X1) variable on the Purchase Intention (Y) variable, for Halal Certification (X2) on Purchase Intention (Y) through Brand Image (Z). As a calculation, the relationship between the Halal Certification (X2) and the Purchase Intention (Y) through the Brand Image (Z) showed a two-tailed probability of $0.00001279 < 0.05$, this means that Brand Image (Z) can mediate between the Halal Certification (X2) and Purchase Intention (Y).

## 5 DISCUSSION

A significant relationship between halal awareness, the certification of halal, brand image, and purchase intention were revealed by Aziz and Chok (2013). In this study, respondents gave their respective assessments of the halal certificate. Most consumers prefer to consume halal products and avoided non-halal products. Meanwhile, the guarantee of security for Muslims to use safe products was a halal certificate from the MUI. Thus, consumers will always consider the existence of a halal logo on the product to be purchased.

The image of brands is related to attitudes in beliefs and references to certain brands (Dianita & Arifin 2018). The data obtained is known that respondents have a good assessment regarding the quality, history, and performance of the toiletry's product with the halal logo. Based on the analysis that has been presented, the respondents gave their respective assessments related to trends, recommendations, and purchase intentions. In contrast, most respondents agreed to buy and recommend halal toiletries to others.

Halal awareness can be perceived as a process of getting information to increase understanding of what Muslims can consume (Ambali & Bakar 2014). It can be defined as how much consumers were aware of or understand halal itself, and also their understanding of what brand they will choose. It can be interpreted that halal awareness is almost the same as brand awareness. Based on the same halal awareness and brand awareness, halal awareness is the level of consumer understanding regarding the concept of halal itself and brand awareness is the consumer understanding of the brand image. Apart from Halal Awareness, consumers also consider the brand image of a product. This is because a company's brand image can influence long-term profits, encourage consumers' desire to buy products at affordable prices, increase the selling price of shares, competitive advantages and marketing success (Yoo & Donthu 2001).

According to Setiadi (2003), consumers who have a positive image of a brand were more likely to make purchases. As a result, brand image can influence consumer purchase intention. As proved by Haryoko and Ali (2018), the brand image has a positive and significant effect on purchase intention. With a halal logo on a product, it can be indicated that the product is under halal standards, which also affects the image of a product. This research is in line with the previous study by Wanda (2019) who stated that the halal label and product quality affect the brand image.

# 6 CONCLUSION

In conclusion, this study proved that there is an impact of halal awareness and certificate halal on brand image and purchase intention. Consequently, the findings of this study are important for marketers of halal products in Indonesia and other nations. Marketers will influence and increase customer buying interest by enhancing marketing efforts by putting more attention to these three factors: halal awareness, the halal logo in the product and the brand image. Besides that, the results of this study can be used as a reference for further research because the variables used in this study have been proven to influence each other. Therefore, it is hoped that further research considers other variables to obtain a more varied result and enrich the existing theory.

## REFERENCES

Ambali & Bakar. 2014. People's Awareness on Halal Foods and Products: Potential Issues for Policy-Makers. Procedia-Socialand Behavioral Sciences 121. 3–25.

Aziz & Chok. 2013. The Role of Halal Awareness, Halal Certificaton, and Marketing Components in Determining Halal Purchase Intention Amon Non-Muslim in Malaysia: A Structural Equation Modeling Approach. Journal of International Food & Agribusiness Marketing, 25, ISSN: 0897-4438, 1–23.

Borzooei & Asgari. 2013. The Halal Brand Personality and Its Effect on Purchase Intention. Interdisplinary Journal Of Contemporary Research In Business Vol. 5, No. 3, 481–491.

Dianita & Arifin. 2018. Pengaruh Brand Ekstensi Dan Brand Image Terhadap Minat Beli Iphone Di Kota Malang. Jurnal Administrasi Bisnis (JAB) Vol 64, No. 1. 119–125.

Endah, N. H. 2014. Consumers's Purchasing Behavior Toward Halal Labeled. Jurnal Ekonomi dan Pembangunan Vol. 22, No. 1. 11–25.

Fitria et al. 2019. The Effect Of Halal Awareness, Halal Certification and Halal Marketing Toward Halal Purchase Intention Of Fast Food Among Muslim Millenials Generation. RJOAS, 6(90), 76–83.

Hidayat & Resticha. 2019. Analisis Pengaruh Variasi Produk dan Labelisasi Halal Terhadap Kepuasan Konsumen untuk Meningkatkan Minat Beli Ulang Pada Kosmetik Wardah. Journal of Business Administration Vol. 3, No. 1, 40–52.

Izzuddin, A. 2018. The Effect Of Halal Labels, Halal Awareness and Food Materials On Interest To Buy Culinary Foods. Jurnal Penelitian Ipteks Vol. 3, No. 2. 100–114.

Janah, M. 2018. Pengaruh Kesadaran Halal Dan Sertifikasi Halal Terhadap Minat Beli Produk Mi Samyang. Journal Management. 1–12.

Majid et al. 2015. Consumer Purchase Intention towards Halal Cosmetics & Personal Care Products in Pakistan. Global Journal of Research in Business & Management Vol. 1, No. 1. 45–53.

Nugraha et al. 2017. Pengaruh Labelisasi Halal Terhadap Minat Beli Konsumen (Survei Pada Mahasiswa Muslim Konsumen Mie Samyang Berlogo Halal Korean Muslim Federation Di Kota Malang).Jurnal Administrasi Bisnis (JAB) Vol. 50, No. 5. 113–120.

Nurcahyo & Hudrasyah. 2017. The Influence Of Halal Awareness, Halal Certification, and Personal Societal Perception Toward Purchase Intention. Journal Of Business and Management Vol. 6, No. 1, 21–31.

Setianingsih & Marwansyah. 2019. The Effect of Halal Certification and Halal Awareness through Interest in Decisions on Buying Halal Food Products. Journal of Islamic Economics, Finance and Banking Vol. 3, No. 1. 64–79.

Shaari & Arifin. 2010. Dimension of Halal Purchase Intention: A Preliminary Study. International Review of Business Research Papers, 6(4), 1–15.

Shukla, P. 2010. Impact of Interpersonal Influences, Brand Origin and Brand Image on Luxury Purchase Intentions: Measuring Interfunctional Interactions and A Cross-National Comparison. Journal of World Business 46, 242–252.

Sumartik et al. 2019. The Influense of Halal Labelization, Brand Image and product Quality Towards Consumer Purchase Decisions of Wardah Cosmetic Product. International Conference on Economis, Management, and Accounting. 13–21.

Tim Global Pathfinder Report: Halal Trends Agriculture and Agri-Food Canada. 2011. http://www.gov.mb.ca/agriculture/market-prices-and-statistics/food-andvalue-added-agriculture- statistics/pubs/halal_market_pathfinder_en/. (Online) accessed on October 1, 2019.

Tim Tribunnews. 2016. Permintaan Pasar Global Terhadap Produk Halal Meningkat, Peluang Buat Indonesia dan Malaysia. https://www.tribunnews.com/bisnis/2016/08/02/permintaan-pasar-global-terhadap-produk-halal-meningkat-peluang-buat-indonesia-dan-malaysia/. (Online) accessed on October 1, 2019.

Tim World Halal Forum. 2013.http://www.worldhalalforum.org/whf_intro.html/. (Online) accessed on October 1, 2019.

Undang-Undang No 33 Tahun 2014 Tentang Jaminan Produk Halal.

Widyaningrum, P. W. 2019. Pengaruh Label Halal, Kesadaran Halal, Iklan dan Celebrity Endorser terhadap Minat Pembelian Kosmetik melalui Variabel Persepsi sebagai Mediasi. Jurnal Ekonomi dan Manajemen Vol. 2, No. 2. 74–97.

*Halal Development: Trends, Opportunities and Challenges – Pratikto et al (eds)*
*© 2021 The Author(s), ISBN 978-1-032-03830-8*

# Exploring perception and resistance on halal labeling: An insight from Denpasar, Bali

L.A. Perguna*, S. Triharini & D.S. Pahlevi
*Universitas Negeri Malang, Malang, Indonesia*

ABSTRACT: This article aims to describe the dynamics of halal labeling in Bali, especially in Denpasar City. Halal dynamics have become a paradox within the non-Muslim community. This research used a qualitative approach. Data was gathered through interviews and non-participative observation of food stall owners, Denpasar's Hindu indigenous and Muslim residents, selected by combining purposive and snowball methods. The first group interpreted the existence of halal food stalls as enriching food choices. Halal food stalls influence Bali's social economy. The second group was resistant to halal labeling. The opposition was showed by introducing halal's anti-thesis, which is called *Sukla*. The third group is moderate. This group presumed that halal labeling strengthens tourism destination image. However, it might slowly deface Bali's position as a cultural tourism destination.

*Keywords*: halal labeling, dynamics, significant, informal sector, *Sukla*

## 1 INTRODUCTION

The great potential of global Muslim tourists visiting Indonesia has made halal tourism one of Indonesia's halal industry's supporting sectors. The halal industry is one of the fastest-growing business sectors in the global market, which includes finance, service, transportation, tourism, and food. In 2019, Indonesia, along with Malaysia, placed on top as countries with the best halal tourism destination. There are many factors of why Indonesia placed top as having the best halal tourism, one of which was halal dishes and food availability that are widespread and easy to find.

Food is a necessity of life that provides energy and nutrition for growth, and there is tremendous market potential for halal food as Islam is also growing at a fast pace (Khalek & Ismail 2015; Munich Personal RePEc Archive (MPRA) 2016). Many food companies are applying for halal certification to meet the demand, especially in European countries. The availability of halal dishes, especially halal-certified dishes, is an important aspect of Muslims' traveling (Dinar standard 2012). Halal certification or halal labeling is a benchmark for Muslim tourists that the food and beverages they consumed are guaranteed to be halal because of the ingredients, agriculture, management, and presentation (Jaelani 2017; Soon et al. 2017).

Currently, food has become the first priority halal lifestyle sector among Muslims. The need for halal food has turned it into a halal lifestyle. The halal lifestyle makes a potential market to be developed from upstream to downstream industries. One of the industries is the tourism industry. As part of the industry, tourist destinations must provide halal services ranging from products, services, and facilities to the environment (Vargas-Sánchez & Moral-Moral 2019). The availability of halal services is essential as an effort to maintain a destination place image.

One of the areas that has become a reference for tourist destinations in Indonesia is Bali. In Bali, especially Denpasar city, halal services have started to disseminate massively in several attributes, including food, beverage, and restaurant products. Food stall services are easy to find in Denpasar

---

*Corresponding author

DOI 10.1201/9781003189282-11

city. The stalls can survive and continue to expand amid the major Hindu society of Denpasar. Halal labeling in food stalls or restaurants in tourism is an important element for domestic and foreign Muslim tourists and the indigenous Muslim community in Denpasar, which amounts to 28% of the total population (BPS 2010). For Muslims, halal labeling is very important when looking for a culinary place, whether it is officially labeled or only by the seller. Halal discussion is related to food products and non-food products such as cosmetics, pharmaceuticals (medicines), leather crafts, fragrances, and services such as banking, entertainment, tourism, and logistics (Rahim & Shahwan 2013). The widespread aspects of halal are the result of the elaborated definition of the word halal, which is often constructed differently (De Nastiti & Perguna 2020). This halal construction has implicated the public's response to the word halal. On the other hand, the word halal is sometimes counterproductive to other cultural values (Wilson & Liu 2010). The norms of behavior and teachings of Islam are often imagined to contradict Western societies' norms or other non-Muslim societies (Henderson 2006). For example, some religions allow the consumption of pork, but it is prohibited for Muslims. This is understandable as the strength of culture and religion greatly influences an individual's perceptions and behavior towards themselves and others (Delener 1994; Essoo & Dibb 2004; Zamani-Farahani & Musa 2012).

On the other hand, the word halal can also have a positive definition, especially in the context of fulfilling Muslims' needs, boosting regional economies, expanding consumers' market segmentation, especially in the field of culinary tourism, and gradually changing the image of Bali as 'non-halal' tourist destination (Fischer 2011; Republika 2015). The massive presence of halal stalls in Denpasar is a sign that halal can have a positive meaning. Much research on halal tourism focuses on destination marketing and economic issues (Battour & Ismail 2016; Bttour et al. 2010a; Carboni et al. 2014; El-Gohary 2016; Henderson 2016; Khan & Callanan 2017; Ryan 2016). Perceptions about halal tourism and result products have been discussed. However, publications discussing halal labeling dynamics and its exploration of stalls as the informal sector in Muslim minority areas are yet to be available. This article focuses on halal labeling dynamics in the informal sector such as stalls in Muslim minority areas of Denpasar city.

## 2 METHODS

Exploratory qualitative research was used in this study. An exploratory study was conducted to explore the dynamics of informal sector-based halal restaurants or stalls in majority Hindu believer's locations. The research was in Denpasar city. As the capital city of Bali Province, Denpasar can represent the condition of Bali, although not to generalize other cities/regencies in Bali. Data collection techniques were carried out through observations and interviews with halal food stall owners, both those with halal and non-halal labels. Interviews were also conducted with Denpasar's Hindu indigenous (former Regional Representative of the Republic of Indonesia candidate) and Denpasar's Muslim residents (Indonesian Ulema Council and local mosque organization figures). The research was conducted from July to September of 2020. The informants were selected by using purposive and snowball techniques. This technique was carried out so that the data obtained were representative and reliable to explore the dynamics of halal culinary entrepreneurs in Denpasar who developed their business as an integral part of tourism in Bali. The focus of the question is the causal effect of halal labeling, the response and resistance toward the labeling, and the image of Bali as a tourist destination after the expansion of halal labels.

## 3 RESULTS AND DISCUSSION

### 3.1 Halal labeling as signification

Many businesses and entrepreneurs do not realize that having a halal-certified businesses implies a big responsibility towards Muslim consumers (Ibrahim et al. 2010). Especially in Indonesia, where more than 80% of the population is Muslim, the obligation to even need a halal certificate is still quite low, though halal certification is an administrative and legal proof of the halal-ness of food products produced by the culinary business. On the one hand, culinary businesses' existence continues to flourish in Denpasar from large to small, both modern and traditional. On the other hand, if the focus is on halal-certified food availability, it is still difficult to find food that is officially halal-certified by the Indonesian Ulema Council as the official institution that issues the certificate. Only 3% of dining or restaurants are halal certified throughout Indonesia (Paramitha 2018).

There are many obstacles and challenges faced by dining places and restaurants to have an official halal certificate, including in Denpasar. Problems that arise are due to limited funds, time-consuming processes, and limited knowledge, especially for the informal sector (food stalls in the context of this article). The status of halal food stalls included in the informal sector category causes them not to consider the necessity for an official halal certificate. For them, the most important thing is showing the word halal or Muslim. These small-scale actors assumed that Indonesia is a Muslim country, so self-labeled halal is enough to automatically indicate their food as halal. Halal became its own identification, which shows that consumers can find halal food and beverages that don't have a halal certificate. The inscription is a signifier for the consumers. A signifier is a meaningful streak that is captured by the senses physically and is very materialistic. Because of the importance of halal or Muslim inscription, almost all food stalls in Denpasar uses this label. More than 600 food stalls in Denpasar uses this signifier. Their reason is simple; to indicate the stalls are serving food that can be consumed by Muslims. At the same time, it is part of a marketing strategy to attract non-Muslim consumers with no food prohibition under religious teachings outside Islam.

From the existence of the signifiers "halal" and "Muslim", the concept of the signified is raised. The signified is a conception or symbol of the signifier. Halal labels that appear in food stalls, although not officially interpreted by consumers, are stalls that provide food that is allowed and permitted to be consumed according to Islamic values. There are no prohibited components that cover all aspects, starting from agriculture to food on the table (Soon et al. 2017). The relationship between the signifier and signified, which cannot be separated, is known as significant. Halal is a significant concept. In the significant concept, the signifier and the signified relationship are arbitrary as various signifiers can be used to show the same concept.

### 3.2  Halal foods stalls dynamics

One of the fast-growing informal sectors in Denpasar is the food stall business. As far as the eye can see, we can easily find food stalls. The existence of these stalls has an important role in the effort to contribute to urban development as it can occupy labor and contribute to the government's revenue. These stalls also contribute to supporting tourism by providing facilities and services for tourists in Denpasar. This facility is clearly needed, especially for Muslims, to give easy access to a halal place to eat. Halal food is a priority and Muslim's main lifestyle when deciding whether to eat or drink. They will try hard to find a halal place to eat in Denpasar, especially for visiting Muslim tourists. As Muslims are a minority, both tourists and Muslim residents in Denpasar will not be careless in choosing where to eat. They try hard to avoid foods that are prohibited by Islamic law in Denpasar, such as pork, dog meat, or others. The halal life-style is mainly on food, which in turn also causes the increase of halal labeled food stalls. The number exceeds restaurants or stalls that offered pork, dog meat, and prohibited things by Islamic law. Based on census-based research results, there are 616 food stalls dedicated to Muslims and 123 restaurants/food stalls exclusively for non-Muslims. This data shows the existence and acceptance of halal food stalls in Denpasar. This number also shows that halal products and services do not exclusively belong to the Muslim market but also all societies.

In Denpasar, there are three typologies of society regarding the massive presence of halal food stalls. The first type is a group of people who agree with the presence of halal food stalls. Denpasar residents (non-Muslims) think that halal food stalls provide a choice for the community other than foods that are popular in Bali but forbidden for Muslims such as *Babi guling* (roasted pork), *sate plecing* from pork, *lawar* (bats), and others. The group that agrees considered halal food to have a distinctive taste and considered it different from non-halal food (made from pork). Halal food products are considered attractive to non-Muslim consumers due to safety, hygiene, and health concerns (Stephenson 2014). These halal dining place or restaurants are inclusive for anyone. This inclusiveness is also advanced by some sellers in halal food stalls by not selling products made from beef. The cow is considered a noble animal by Hindus as it is the vehicle of Lord Shiva. Due to its glorification, Hindus do not eat beef. This is different from the Islamic concept of not eating pork because it violates Islamic law (haram). As for the Muslim community, the presence of many halal stalls in Denpasar is their main choice. This choice is based on the Muslim belief that eating halal food is a religious command. They will never try foods other than those with a halal label, although there are Muslims who still consume non-halal food.

A second group is a group that is resistant to the presence of halal food stalls in Denpasar. The forms of opposition vary. For the seller, they put the label 100% Haram. This label is the anti-thesis of 100% Halal. Muslim or first-time Muslim tourists will be surprised to see the label. This phenomenon occurred for the first time since 2014 as a response to the issuance of Law Number 33 of 2014 concerning the Halal Product Warrant. This law requires all products to be halal certified. This certificate will be issued by the Indonesian Ulema Council when the product is declared halal. Halal means fit for consumption and under Islamic provisions. Meanwhile, for the Balinese, who are predominantly Hindu, their merchandise is processed food from pork, which in Islam is prohibited. Even though they sell food made from halal ingredients, it is not certain that they pass the halal certification because the cooking method is different and may not be considered under Islamic law. This was the situation of a 100% haram label for culinary entrepreneurs in Bali in general. The act was a form of protest against the law because it is considered as the domination of Islam as the majority religion in Indonesia, even though Indonesia is not a country with Islamic principles.

Another form of resistance was that several community groups created social movements with the big theme of *Sukla Satyanugraha*. This movement is a form of dominant resistance against the regulation. Although Sukla's use term as food that is suitable for consumption by Hindus is still a polemic, this movement voices and urges Balinese people to shop at Balinese stalls and not shop at stalls owned by Muslims or outside Bali. Another extreme resistance measure is building a certification movement for Bali's food stalls with the label *Sukla*. *Sukla* does not mean halal or non-halal (Haram). In simple terms, *Sukla* means the process of purification of food from preparation to serving. This movement is considered able to unite Hindus amid the hegemony controlled by the newcomer. The researchers observed that newcomers living in Denpasar, some of whom are Muslims, are reluctant to shop for food owned by the Balinese. On the other hand, the native Balinese do not have problems shopping and buying food at stalls owned by newcomers. This condition benefits newcomers' traders as their customers are not only from the Balinese Hindu community but also from the Muslim community. As an affirmation and confirmation of identity, Muslim food vendors put the words 100% halal, *Warung Muslim*, *Warung Jawa* and others. This phenomenon causes the Balinese to feel recessive that resulting in the creation of several phrases, such as *Orang Bali Jual Tanah Untuk Beli Bakso* (Balinese sells land to buy meatballs), *Jawa Jual Bakso untuk Beli Tanah* (Javanese sells meatballs to buy land).

The third group is neutral and impartial with many halal stalls in Denpasar. This group can be called the moderate group. This group assumes the halal labeling was not made following authority interests through the regulation establishment, but rather following the necessity of the consumers, including consumers in Bali who are mostly tourists. The non-Muslim Denpasar community understands the vast amount of food stalls that are labeled halal. This is seen as an effort to reflect Denpasar and Bali as friendly to Muslim tourists. Apart from the need to fulfill the necessity and the image of Denpasar tourist destinations, for sellers, the emergence of halal labeling is considered a marketing strategy to attract consumers, especially Muslims. Since the need for halal food has become a Muslim lifestyle, it is only natural for destinations to provide this halal food with the available food stalls. When a tourist destination such as Denpasar provides halal products, it will impact tourist visits, especially Muslims, to these tourist destinations. This tourist visit will provide a multiplier effect for tourist visiting areas.

They also viewed that if the *Sukla* labeling initiated by the *Sukla Satyanugraha* group could provide a sense of comfort for consumers, it would certainly invite consumers to come shopping. The nature of halal and *Sukla* labeling is a matter of each individual's belief, so it is very basic and personal. The moderate group also thinks that Denpasar's vast halal label will slowly eliminate and marr Bali's image as a cultural tourism destination with its own local wisdom. This group does not show open resistance like the second group. However, they are more focused on how this halal-labeled shop can flourish in Denpasar and how indigenous (Hindu) stalls can also survive and develop. They assume that the presence of this shop can also provide comfort for Muslim tourists, especially in carrying out worshipping under the Islamic law when traveling without intending to alienate and limit non-Muslim tourists from one tourist destination. It is hoped that halal stalls services can provide safe and healthy services while introducing Islamic teachings without discriminating against other teachings.

## 4 CONCLUSION

The halal label is the most important aspect in food/beverages along with the increasing of halal lifestyle in Indonesia, including in Denpasar, Bali as one of the most popular tourist destinations in Indonesia. For the image of tourist destinations to continue to increase, support from all elements is needed, including halal food stall services. Due to its importance, this halal label can be observed to be flourishing of the number of food stalls with a halal label in Denpasar, which is four times more than the number of non-halal food stalls. Its existence triggers a dynamic in the community. Some agree with halal labeling. Some strongly oppose and against the halal label, namely *Sukla*. Some are neutral with a variety of rationales.

REFERENCES

Battour, M., & Ismail, M. N. 2016. Halal tourism: Concepts, practises, challenges and future. *Tourism Management Perspectives*, *19*, 150–154.

Battour, M. M., Ismail, M. N., & Battor, M. 2010a. Toward a halal tourism market. *Tourism Analysis*, *15*(4), 461–47.

Carboni, M., Perelli, C., & Sistu, G. 2014. Is Islamic tourism a viable option for Tunisian tourism? In-sights from Djerba. *Tourism Management Perspectives*, *11*, 1–9.

Delener, N. 1994. Religious contrasts in consumer decision behaviour patterns: Their dimensions and marketing implications. *European Journal of Marketing*.

De Nastiti, N., & Perguna, L. A. 2020. Konstruksi Konsumen Muslim Terhadap Labelling Halal Pada Produk Kosmetik (Studi Fenomenologi Penggunaan Kosmetik Halal Di Kalangan Mahasiswi Di Kota Malang. *Jurnal Analisa Sosiologi*, *9* (1).

El-Gohary, H. 2016. Halal tourism, is it really Halal? *Tourism Management Perspectives*, *19*, 124–130.

Essoo, N., & Dibb, S. 2004. Religious influences on shopping behaviour: An exploratory study. *Journal of Marketing Management*, *20*(7–8), 683–712.

Fischer J. 2011. The Halal Frontier. In: The Halal Frontier. Contemporary Anthropology of Religion. Palgrave Macmillan, New York. https://doi.org/10.1057/9780230119789_1

Henderson, J. C. 2006. Tourism in Dubai: Overcoming barriers to destination development. *International Journal of Tourism Research*, *8*(2), 87–99.

Henderson, J. C. 2016. Halal food, certification and halal tourism: Insights from Malaysia and Singapore. *Tourism Management Perspectives*, *19*, 160–164.

Khan, F., & Callanan, M. 2017. The "Halalification" of tourism. *Journal of Islamic Marketing*.

Mohsin, A., Ramli, N., & Alkhulayfi, B. A. 2016. Halal tourism: Emerging opportunities. *Tourism Management Perspectives*, *19*, 137–143.

Rahim, N. F., & Shahwan, S. 2013. Awareness and perception of muslim consumers on non-food halal product. *Journal of Social and Development Sciences*, *4*(10), 478–487.

Ryan, C. 2016. Halal tourism. *Tourism Management Perspectives*, *19*, 121–123.

Soon, J.M., Chandia, M. and Regenstein, J.M. 2017. "Halal integrity in the food supply chain", *British Food Journal*, Vol. 119 No. 1, pp. 39–51.

Stephenson, M. L. 2014. Deciphering 'Islamic hospitality': Developments, challenges and opportunities. *Tourism Management*, *40*, 155–164.

Vargas-Sánchez, A., & Moral-Moral, M. 2019. Halal tourism: State of the art. *Tourism Review*.

Wilson, J. A., & Liu, J. 2010. Shaping the halal into a brand? *Journal of Islamic Marketing*.

Zamani-Farahani, H., & Musa, G. 2012. The relationship between Islamic religiosity and residents' perceptions of socio-cultural impacts of tourism in Iran: Case studies of Sare'in and Masooleh. *Tourism Management*, *33*(4), 802–814.

*Halal Development: Trends, Opportunities and Challenges – Pratikto et al (eds)*
*© 2021 The Author(s), ISBN 978-1-032-03830-8*

# Is halal certificate socialization effective in increasing the number of MSMEs in the food sector to register for halal certificates?

M. Puspaningtyas
*Universitas Negeri Malang, Malang, Indonesia*

ABSTRACT:   The purpose of this research is to provide knowledge and application to The Owner of SMEs to register their products at the health office so as to get halal labels. Halal food is an important requirement for the advancement of local food products in Indonesia to compete with other products both at home and abroad. Indonesia is a country with a majority Muslim population. Based on the presentation of the results and discussion above, it can be concluded from the service activity that awareness exists of the importance of the ownership of letters, which significantly supports business strengthening. Business actors are very enthusiastic about participating in counseling activities about the ownership of halal certificates.

*Keywords*:   halal, register, small-medium-enterprise

## 1   INTRODUCTION

Indonesia is a country with many Islamic religions that dominate the religious population statistics. 87.2% or as many as 207 million people are Muslims (Indonesia.go.id). For Muslims, it is essential to know whether the food is lawful or not. Because according to Islamic law, the importance of food should be considered. As it appears in the Holy Qur'an the letter of Al-Baqarah (168–171) reads: "O people eat what is clean and good that is on the earth, and do not follow in the steps of Satan. Because indeed Satan is a real enemy to you. (169) Indeed, the devil only orders you to do evil and wickedness and tells Allah what you do not know. (170) And when it is said to them: 'Follow what Allah has sent down,' and they answer: '(No), but we only follow what we have found from the (deeds of) our ancestors.' '(Will they follow too), even though their ancestors did not know anything, and didn't get a clue?' (171) And the parable (of the one calling) the disbelievers is like the shepherd who calls the beast who hears nothing but calls and cries. They are deaf, dumb, and blind, so (therefore) they don't understand." (Al-Baqarah, 168–171).

Fulfillment of halal guarantees must be made to meet food production standards while protecting consumers. Etymologically, halal is known as "*halla*" which means "off" or "not tied". *Halalan* is anything that is allowed and is not bound by various provisions. *Tayyib* means "delicious, good, healthy, reassuring". Meanwhile, if the food is good and not dirty, both from the substances it contains and is not mixed with unclean objects, it is known as *tayyib* (Girindra 2008). According to Regulation of Constitutions number 8, the year 1999 concerning Consumer Protection explains consumer protection exists to provide benefits, fairness, balance, security and consumer safety, and legal certainty. Furthermore, Regulation of Constitutions number 7, the year 1996, concerning Food, article 30 paragraph (1) and (2) stipulates that every person who manufactures or imports Indonesian food packaged for trade must include a halal label so that consumers avoid consuming non-halal products. The purpose of enacting the necessity of food entrepreneurs to give a halal label on each product is to help increase consumer confidence in the quality of the food product. When consumers have given their trust in a product, consumers will tend to choose products that are believed to be of quality. Indirectly, sales of Micro, Small, and Medium Enterprise (MSME) products will increase and grow. Hard work is needed in realizing a thriving business. Various kinds of efforts need to be made by business owners to promote their products.

According to the Chairperson of the Indonesian Ulama Council (MUI), KH. Ma'ruf Amin in opening the 2018 International Indonesia Halal Expo (INDHEX) explained that halal certification becomes the strength of small-medium-enterprise competitiveness because consumer demands for halal food availability are increasing. Halal certification becomes important for food producers due to several things, such as guaranteed halal food products, the product is more trusted by the people, benefit from the marketing side, the more convincing the consumer, considers it too important for halal not located on the label. While food manufacturer information on halal certification is obtained using several ways, such as print and electronic media, BPOM, Ministry of Health, seminars, relatives or friends, and based on the labels on the packaging. The question that arises over the many socializations that are often carried out is "how much of the socialization's effectiveness is required for the realization of MSME actors to register themselves to get a halal certificate for their business?"

The reality shows that there are so many new Micro, Small, and Medium Enterprises (MSME) emerging every year. The latest data from www.dekop.go.id (2019) shows an increase of 2.02% in the 2017–2018 timeframe. However, the increase in the number of SMEs is still far behind the number of halal-certified products. Data obtained from halalmui.org states that in 2018 there were only 204,222 products and 11,249 companies that were halal certified. In 2019, the number of MSMEs with halal certification increased by 274,796 products and 13,951 companies (halalmui.org 2019). Unfortunately, this increase has not shown a significant value because when compared to the number of Micro, Small, and Medium Enterprise (MSME), the percentage is very small each year.

Based on the description above, we need an activity that is right on target to create food that is safe and does not make Muslims still wonder about the halal nature of the food. Thus, this community service activity with the theme of Halal Certificate Education and P-IRT for Food Products at SMEs throughout Blitar Raya. This community service activity aims to motivate the community of entrepreneurs to have a strong foundation in running their business. This activity is carried out to socialize the importance of the P-IRT number in ensuring the business's continuity and quality and the benefits of halal certificates as the selling power of the product. From the academic perspectives, this study may contribute to other researchers as their main references regarding the study on educating SMEs about Halal certification's importance.

## 2 LITERATURE REVIEW

Halal products contain goods or services related to food, beverages, medicines, cosmetic chemical products, or biological products, as well as users of goods that are used by the community. The halal product process is a series of activities to ensure that a product comes from providing materials, processing, storing, packaging distribution, selling, and presenting the product in a good way. The guarantee of a halal product requires a system that contains halal assurance, both in terms of raw materials and their derivatives as well as from the production process. Foods that are fit for consumption consist of quality elements. The quality of food is not just about its looks, texture, and taste. In the halal concept, quality food includes its hygiene and wholesomeness, and these are among the factors lacking in small-medium enterprises especially when they want to adopt halal (Backhouse & Mohamad 2014). The halal guarantee system (SJH) is a system created and implemented by companies holding halal certificates to ensure the halal production process's sustainability. This system is made into part of a stand-alone system's policy so that the resulting product can be guaranteed halal following the rules outlined by LPPOM-MUI (Hasan 2018).

MUI halal certificate is a requirement to obtain permission to put halal labels on product packaging from authorized government agencies. Producers guarantee the halal production process's continuity by implementing the Halal Assurance System (www.halalmui.org). The halal certification process is a series of procedures for obtaining halal certificates for products to be marketed. In line with the statement, some scholars stated that the foodservice industry had some challenges in halal food certification (Kurth & Glasbergen 2017). Among the challenges include cost (Rahman et al. 2016), lack of government support (Shafie & Othman 2006), and awareness (Ngah et al. 2015). However, according to Salikin et al. (2014), adopting a quality certification requires more than cost, support, and awareness. It requires motivation (Koryak et al. 2015) and knowledge.

Additionally, many studies have focused on halal awareness and perception (Elias et al. 2016), halal certification (Sulaiman et al. 2016), and halal supply chain (Soon et al. 2017). Thus, knowledge in the motivational factors influencing the adoption of halal certification in the small and micro-enterprise are still scarce. This study could improve knowledge and help the government in introducing a policy to encourage these companies to embark on the halal journey (Basir et al. 2018). Halal products can convince consumers to buy a product so that sales made by business owners experience an increase. Also, halal certification is a potentially useful branding and marketing mechanism for products and services. By understanding the factors on halal certification adoption for small and micro enterprises, the policymaker may know the level of acceptance on halal certification adoption for the small- and micro-enterprises. Some of Indonesia's development planning and direction documents have explicitly stated efforts to strengthen small businesses. Azqiara (2019) also argued that MSMEs could be interpreted as the development of a mainstay area to accelerate economic recovery to accommodate priority programs and development of various sectors and potentials, while small businesses are an increase in various efforts of community empowerment. There are groups of actors in the upstream-downstream chain who cannot accumulate capital in the four research cases. These groups include ceramic craftsmen and coconut sugar groups. Meanwhile, in the furniture industry, the second-level subcontracting relationship with the labor groups cannot accumulate capital. Another group that was found unable to raise capital was the rattan picking farmer group (Widyaningrum 2013).

## 3 METHODS

This research employed a qualitative approach. The implementation of preliminary activities in socialization activities was designed with P-IRT education activities and halal certification by the Indonesian Ulema Council in the Blitar Regency. In addition to the education process for MSME owners, this service activity also assists entrepreneurs who do not have a halal certificates in obtaining a halal certification.

Data collection methods used interviews and observations of 33 subjects who participated in the socialization of the importance of halal certificates. A total of 33 participants were asked the same questions in the context of the benefits of the outreach activity. The selection of 33 participants was tentative, meaning that it is adjusted to the conditions of MSMEs in Blitar City. The sampling technique used was snowball sampling, so that not all participants were interviewed when the answers submitted were relatively the same. We also conducted participant observations on socialization activities in community service activities about enthusiastically registering for halal certificates through questionnaires.

Data reduction was based on business owners' enthusiasm to register for a halal certificate after the socialization activity took place, evidenced by filling in the Halal Certificate Registration Form. The selected samples were 33 MSME owners. Business owners were interviewed with structured questions to determine the effectiveness of the previous socialization. The interview technique used snowball sampling to end the interview, and the data was deemed sufficient when the answers given by the research subject were saturated or were the same.

## 4 RESULTS AND DISCUSSION

### 4.1 *The motivation aspect*

Based on the community's problems regarding the effectiveness of the socialization of the MUI halal certificate for MSMEs, the enthusiasm of the owners of MSMEs in obtaining halal certificates is very good. It is shown from the results of the questionnaire filled out by 33 research subjects who are willing to fill out the MUI halal certificate registration form. Previous research by Agustina et al. (2019) revealed that, after the socialization in the form of counseling, SME players had a deeper understanding of the importance of halal certification in their food products, increased insight into the effect of halal certification on buyer satisfaction of their products, created prospects business, as well as obtained the formation of understanding and experience regarding the steps for applying for halal certification to The Institute for the Study of Food, Drugs and Cosmetics of the Indonesian Ulema Council or LPPOM MUI using Cerol services.

The counseling has included the latest information about consumer behavior and consumer trends in choosing the type of food product. With halal certification, it is expected to increase competitiveness among similar industries. Halal certified products can ensure that the products have good nutritional content, are hygienic, healthy, and do not contain ingredients that can endanger the health of consumers. In line with that, Basir et al. (2018) conclude about his Based on previous studies and findings, expected competitiveness can influence the food safety management system adoption (Rahman et al. 2016). The findings from Rahman et al. (2016) stated that when the managers considered and realized the sources of competitiveness, it resulted in the adoption of a food safety management system. Also, Fernando and Yusoff (2014) highlighted that top management support positively impacts the Malaysian food safety system's adoption. In line with that, Rahman et al. (2016) contended that top management support is a crucial factor in the adoption of any management system. For instance, Hamim et al. (2015) agreed that top management commitment is a critical success factor of quality management practice. Arpanutud et al. (2009) stated that top management commitment would influence food safety management systems' implementation. To confirm this (Arpanutud et al. 2009) also stated that top management in the firm will increase food safety management system adoption.

Based on the survey in a group of SME in Blitar, the owner of the SME's commitment is significant to halal certification adoption. The researchers concluded that top management plays an important role in ensuring that halal certification adoption works smoothly with many perspectives of the study and variety of research settings. Motivational factors are vital in halal certification due the commitment of higher management. By showing their commitment to halal certification adoption, the process of adoption is run more smoothly as management may provide time, money, and ideas in the adoption of the new policy. In inspiring the halal certification adoption, the study on the motivational factors is important because it provides significance to researchers, academics, and the industry by providing insight into the halal quality certification. Also, the study will add to the body of knowledge. Also, this paper provides valuable information for SME companies in Malaysia to adopt halal certification. The research project is expected to be a reference for future researchers and valuable in directing their focus, efforts, time, and resources on halal certification adoption. The community service activities explored the importance of P-IRT registration and ownership of halal certificates for food entrepreneurs to strengthen micro-businesses' existence. Through this educational activity, it is expected that more micro-business products have been registered with the health department and have a halal label so that the possibility to compete with products that have gone through the manufacturing process is very large. P-IRT can assist SMEs in matters of legal certainty and certainty about the quality of food products.

## 4.2  *Halal certificate registration constraints*

According to Maryati et al. (2016), the process of applying for a halal certificate is based on the current provisions of the LPPOM-MUI. Business actors must understand the requirements for halal certification and attend training in the halal assurance system and apply with a complete set of documents that includes: a list of products, a list of materials, and material documents; a product matrix, a manual of the halal assurance system, a diagrams process flow, an address list of production facilities; evidence of: halal policy dissemination, internal training, an internal audit, registeration for halal certification, monitoring pre-audits and payment of certification contracts, conducting audits, monitoring post-audits, having P-IRT/MD distribution permits, and possession of an MD distribution license for high-risk products. These requirements can be prepared by the business actor except the MD distribution permit from BPOM and IUI/IUMK from BPPTPM.

The long process of registering a halal certificate for a product at MUI has discouraged business owners. In fact, there are other obstacles faced by business owners, such as the costs incurred to prepare halal certificate registration documents. However, a very influential obstacle for SMEs involved in the Food and Beverages sector in registering for a halal certificate is the matter of halal raw materials made from animal elements. According to Maryati et al. (2016), the marketing authorizations problem in halal certificate registration is improving the competitiveness (0.575) with assistance (0.321) from *Badan Pengawas Obat dan Makanan* (0.484). Factors that mostly influence the production facility (0.572) included raw material (0.233), the attitude of enterprises (0.838), and legality (0.432). The best alternative strategies recommended are fostering good production (0.343), bureaucratic of marketing authorization accelerated and simplified (0.169), and intensive coaching (0.153).

## 5  CONCLUSION

This research has uncovered the awareness of the importance of the ownership of letters, which significantly supports business strengthening. Business actors are very enthusiastic about participating in counseling activities about the ownership of halal certificates. This enthusiasm was evidenced by the number of participants who shared their obstacles and problems with the speakers. Apart from sharing activities, participants also asked a lot about the correct management procedures for halal certificates. On the other hand, the socialization's effectiveness, which shows good results, is inversely proportional to post-activity in the field, which has obstacles because of unresolved technical issues.

## REFERENCES

Ade Rosilawati. 2011. "Pengaruh perkembangan Usaha Kecil dan Menengah terhadap Pertumbuhan Ekonomi pada Sektor UKM di Indonesia", Undergraduate Thesis, Yogyakarta.
Aisjah Girindra. 2008. "Dari sertifikasi Menuju Labelisasi Halal." Jakarta: Pustaka Jurnal Halal.
Al-Qur'an, Surat Al-Baqarah (2) ayat 168–171. Al-Qur'an dan Terjemahan. Cetakan ke 7: Al-Mizan Publishing House.
Andriyani. 2019. Kajian Literatur pada Makanan dalam Perspektif Islam dan Kesehatan. Jurnal Kedokteran dan Kesehatan, Vol. 15, No. 2
Arpanutud, P., Keeratipibul, S., Charoensupaya, A., & Taylor, E. 2009. Factors influencing food safety management system adoption in Thai food-manufacturing firms: Model development and testing. British Food Journal, 111(4), 367.
Azqiara. 2019. Pengertian UMKM Secara Umum dan Menurut Para Ahli Lengkap.
Backhouse, C. J., & Mohamad, N. 2014. A framework for the development of Halal food products in Malaysia.
Elias, M.E., Othman, S. N., Yaacob, A. N., & Saifudin, A. M. 2016. A Study of Halal Awareness and Knowledge Among Entrepreneur Undergraduates. International Journal of Supply Chain Management, 5(3), 147–152.
Fernando, Y., Ng, H.H. & Yusoff, Y. 2014. Activities, motives and external factors influencing food safety management system adoption in Malaysia. Food Control, Vol. 41, pp. 69–75.
Kurth, L., & Glasbergen, P. 2017. Serving a heterogeneous Muslim identity? Private governance arrangements of halal food in the Netherlands. Agriculture and Human Values, 34(1), 103–118.
MUI. 2014. Lembaga Pengkajian Pangan Obat-obatan dan kosmetik Majelis Ulama Indonesia.
MUI. INDHEX. 2018, Jalan Menuju Arus Baru Ekonomi Indonesia.
A. H., Zainuddin, Y., & Thurasamy, R. 2015. Barriers and enablers in adopting of Halal warehousing. Journal of Islamic Marketing, 6 (3), 354–376.
MUI. 2019. Data Sertifikasi Halal LPPOM MUI 2012–2019.
Nurul Widyaningrum. 2013. Pola-pola Eksploitasi Terhadap Usaha Kecil. Banding: Aka tiga.
Maria Mohd Salleh. 2018. Motivational Factors for Halal Certification Adoption among Small and Micro Enterprises in Malaysia. Int. J Sup. Chain. Mgt. Vol. 7, No. 4, August 2018. page 391–396.

Rahman, N. A., Yaacob, Z., & Radzi, R. M. 2016. The Challenges Among Malaysian SME: A Theoretical Perspective. World, 6(3), 124–132.

Salikin, N., Ab Wahab, N., & Muhammad, I. 2014. Strengths and Weaknesses among Malaysian SMEs: Financial Management Perspectives. Procedia-Social and Behavioral Sciences, 129, 334–340.

Shafie, S., & Othman, M. N. 2006. Halal Certification: an intern ational marketing issues and challenges. In Proceeding at the International IFSAM VIIIth World Congress (pp. 28–30).

Sofyan Hasan. 2014. "Kepastian Hukum Sertifikasi dan Labelisasi halal Produk Pangan". Jurnal Dinamika hukum, Vol. 14 No. 2.

Soon, J. M., Chandia, M. & Regenstein, J. M. 2017. Halal integrity in the food supply chain. British Food Journal, 119(1), 39–51.

Talib, M. S., Abdul Hamid, A. B., & Ai Chin, T. 2015. Motivations and limitations in implementing Halal food certification: a Pareto analysis. British Food Journal, 117(11), 2664–2705.

Undang-Undang Republik Indonesia Nomor 33 tahun 2014 tentang Jaminan Produk Halal, Jakarta: Hukum Online, 2014.

UU No. 8 Tahun 1999 tentang perlindungan konsumen.

UU No. 7 Tahun 1996 tentang Pangan.

# Potential of halal industry areas to improve national economic growth

N. Muyassaroh* & F. Slamet
*Universitas Negeri Malang, Malang, Indonesia*

M. Sakti
*UPN Veteran Jakarta, Jakarta, Indonesia*

ABSTRACT:   Industrial Revolution Era 4.0 is a new era of industrialization, where Indonesia can further maximize the potential of Natural Resources (NR) and Human Resources (HR) owned. The halal industry is one of the most promising industries in the global market. The large Muslim population in Indonesia is a driving factor that creates great opportunities for the halal industry's growth to improve the national economy. In addition, Indonesia has successfully raised its ranking in the world halal market, ranging from the food sector, Islamic finance, cosmetics, medicines to halal tourism. This shows that Indonesia has great potential to be a major producer and player in the world halal market. Therefore, to realize this target, Indonesia needs to establish a halal industrial area. The concept of a halal industrial area is formed from an existing area. However, with adequate facilities and systems to produce halal products following the Halal Product Guarantee (HPG). The development of this area is intended so that the impact and benefits of the existence of the halal industry can be managed to the maximum with one management group so that it will be more efficient in its operations. Th e urgency of the halal industry area is to maximize the potential of the halal industry as well as the opportunity to become a major producer in the global halal market so as to increase national economy.

*Keywords*:   halal industry, national economic growth, halal revolution

## 1  INTRODUCTION

Industry 4.0 has served as a new round of industrialization that had a big impact on various sectors in Indonesia. Industrialization has transformed Indonesia from an agricultural country into an industrial country contaminated with digitalization. This is evidenced by the government's determination to target Indonesia to become a resilient industrial country by 2035. This determination is motivated by the abundant potential of Indonesia's Natural Resources (NR) and Human Resources (Kemenperin 2017). Not only that, Indonesia's commitment to becoming an Industrial country was even seen before the monetary crisis in 1997-1998. Indonesia was once named as a candidate for an Asian Tiger because of the significant changes in economic structure from an agricultural country to an industry. This is marked using the manufacturing industry as the driving force of national economic progress (Yusof 2011).

Based on data from the Ministry of Industry (2017), non-oil and gas industry growth in the third quarter of 2017 reached 5.49 percent. Non-oil and gas industry branches reached the highest percentage, including the metal industry by 10.60 percent, food and beverage 9.46 percent, and machinery and equipment reached 6.35 percent. However, along with the development of industry in the world, Indonesia has increased in industry in general and in the halal industry. The halal industry market in Indonesia, especially the food, travel, fashion, medicine and halal cosmetics sectors have reached 11 percent in the global market in 2016 (Annisa 2019). The percentage

---

*Corresponding author

DOI 10.1201/9781003189282-13

continues to increase to raise Indonesia's ranking in the Global Islamic Economy Index (GIEI), including Indonesia being ranked first in the Muslim Food category, ranked third in the Muslim Apparel category, ranked fifth in the Muslim Travel category, and ranked sixth in the halal region category (Hakim 2019). These achievements show that Indonesia has great potential to develop the halal industry sector (Annisa 2019). In addition to being influenced by the large Muslim population in Indonesia, this potential is also driven by a global phenomenon that makes the halal market a new growth sector in the global economy, even becoming the most promising sector in Asia, the Middle East, Europe and America. The market for halal-certified food and products is currently growing, both domestically and internationally (Elasrag 2016).

Indonesia is the country with the largest Muslim population. But it is unfortunate that as a country with the largest Muslim population globally, Indonesia is only a destination for halal products from abroad and has not been an important player in the industry as a whole. The government through the National Planning and Development Agency, has inaugurated and launched the Indonesian Sharia Economy Masterplan (MEKSI) 2019–2024. In addition, the National Committee on Islamic Finance (KNKS) was also established to implement MEKSI 2019–2024 into national development and building a halal industry in Indonesia because the development of the Islamic financial economy is not currently in line with expectations. In supporting Sharia economic development through the halal industry, Indonesia is expected to be able to maximize local wisdom in capturing global market opportunities.

The Halal Industrial Estates' existence become very important to facilitate in terms of supervision and monitoring, considering that Indonesia has now become a country that has increased its ranking in the halal industry sector. As the purpose of building a halal industrial area is intended so that the impact and benefits of the existence of the halal industry can be managed to the maximum in one management to be more efficient in its operations (Alrasyid 2016). The development of a Halal Industrial Estates also has the potential to encourage in creased investment in the halal industry sector, provide legal certainty regulate the management of industrial estates (Syahruddin 2010), and increase the demand for goods and services in a region to advance the economy of the community (Istiqomah & Prasetyani 2013). Anchored by this background, this research was designed to explore the potential of halal industrial areas to improve national economic growth.


## 2 LITERATURE REVIEW

Special Economic Zones are areas that have the advantage to maximize industrial activities, exports, imports, and other economic activities that have high economic value. The development of SEZ in Indonesia began with establishing the Free Trade Area and Free Port (FTAFP) or the commonly called Free Trade Zone (FTZ) in 1970. Furthermore, Indonesia designed a Bonded Warehouse Zone in 1972. The next development was followed by the establishment of industrial estates in 1989, Integrated Economic Development Zones (IEDP) in 1996, and most importantly, the Special Economic Zone (SEZ) in 2009 (Suryani & Febriani 2019). As per Law No.39 of 2009, the establishment of KEK aims to accelerate regional development and act as a breakthrough model for regional development for economic growth, including industry, tourism, and trade so as to create jobs (Suheri & Aulia 2017).

One form of SEZ development is the development of industrial estates. This development has also been regulated in Presidential Decree No. 41 of 1996, which was taken to encourage industrial development through industrial sites' construction in the form of industrial estates (Sagala et al. 2004). UNIDO (2012) differentiates the development goals of industrial estates in developed and developing countries. In developed countries it aims to minimize negative externalities (such as pollution and congestion), so that industrial estates are planned as warehousing clusters and distribution centers. In some countries, industrial estates are converted into eco-industrial parks, while in developing countries, industrial estates aim to (1) encourage economic growth and employment at the national, regional and local levels; (2) attract foreign investment; and (3) spur the development of the industrial sector.

The approach to SEZ development must be dynamic in order to improve competitiveness between regions globally. SEZ needs to be oriented towards the competition of science and technology to produce innovations that are the specialty of a SEZ in Indonesia (Suheri & Aulia 2017). A great potential of KEK in Indonesia is the development of Halal Industrial Estates because of the potential of the Indonesian halal industry in the global market today.

## 3 METHODS

This research used a qualitative method with a literature study approach. A qualitative approach is a process of research and understanding based on a methodology that investigates social phenomena and human problems. In this approach, researchers created a complex picture, examined words, detailed reports from respondents' views, and conducted studies on natural situations (Sugiyono 2005). The literature study method included a reassessment of related literature covering systematic descriptions, critical analysis, and the evaluation of relevant texts. The data used for the study of this literature comes from secondary data that includes data released by institutions, papers and journals (Muhyiddin et al. 2017).

## 4 DISCUSSION

### 4.1 *Potential development of halal industry in the global market*

The global halal market has emerged as a new growth sector in the global economy. The halal industry's growth continues to increase in many parts of the world, and the industry is set to become a competitive force in world international trade (Elasrag 2016). This can be proven by achieving the global halal industry estimated to be worth around 2.3 trillion USD (excluding Islamic finance). Growing at an estimated annual rate of 20%, the industry is worth about 560 billion USD per year. This rapid growth is influenced by the size of the world's Muslim population, which is expected to reach 1.8 billion soon. However, with the times, Halal is no longer just a religious obligation for Muslims, but also a strong market force and the market phenomenon around the world for both Muslims and non-Muslims (Azmi et al. 2013).

Many halal industry consumers are increasing the halal market demand, which refers to the needs and desires arising from customer and community pressures (Tieman 2007; Kamaruddin & Jusoff 2009). The halal industry has seen considerable growth and development over the years, driven by increased awareness of halal products among consumers and product manufacturers (Global Leader 2008). Therefore, the halal industry is now expanding outside the food sector, which includes pharmaceuticals, cosmetics, health products, toiletries, and medical devices, as well as components of the service sector such as logistics, marketing, print and electronic media, packaging, branding, and financing (Azmi et al. 2013).

The growth of the halal industry can be seen from the contribution of several halal industry sectors to the world economy. Based on data from The State of the Global Islamic Economy Report 2015–2016, the size and profile of the global Muslim consumer market spending on food and beverages amounts to $1.128 billion (17% of global expenditure). Comparatively, this was a 4.3% increase in Muslim food & beverage spending of $1.081 billion. Muslim food spending is expected to grow to a market of $1.585 billion by 2020 and will account for 16.9% of global expenditure. This equates to a 2014-20 CAGR growth of 5.8% (Elasrag 2016).

In addition to the food and beverage sector, pharmaceutical and health products are also a major growth area in the global halal industry. Demand for pharmaceutical products, generic drugs, health, and halal care are estimated to reach around 555 billion USD in Muslim-majority countries. The growth of the halal medicines markets globally increased by 4% in 2009 to a value exceeding 820 billion USD (Elasrag 2016). Furthermore, in the halal cosmetics sector, the market's growth is reflected by the development of consumer knowledge about the ingredients used, and product awareness, which is driven by social networks. The global halal cosmetics industry is estimated to reach 13 billion USD, with an annual growth rate of 12%. Currently, the halal cosmetics market constitutes 11% of the total global halal industry (Elasrag 2016).

Rapid growth also occurred in the halal tourism sector. The halal tourism market represented 12.3% or 126.1 billion USD of the total global outgoing tourism market and grew by 4.8% compared to the global average of 3.8%. In 2011 alone, Muslim tourists spent around USD126 billion USD. This figure is expected to reach 419 billion USD by 2020. The MENA market represents the largest share of the world's Muslim tourist expenditure by 60%. (Azmi et al. 2013). Based on this phenomenon, it is not surprising that the halal industry trend continues to rise from year to year. Compound Annual Growth Rate (CAGR) projects the halal industry will increase to reach 6.2% in 2018 to 2024 (Fathoni et al. 2020). Of course, this becomes a global market that can continue to be developed in Indonesia.

## 4.2 Opportunities for Indonesia to become a major producer of halal industry

In today's increasingly globalized world, Islamic economies continue to grow and require companies to provide products and services that meet their faith-based needs (Nisha & Iqbal 2017). According to the Global Economic Report, the Sharia economy in Indonesia was ranked 10th in 2018–2019 (Sharia 2019). The ranking is not comparable to Indonesia's Muslim population, reaching 12.6 percent of the total world season population. This figure is equivalent to 87.2% of the total Indonesian population (912 million people) (Hakim 2019). This potential makes Indonesia the largest consumer of the halal industry in the world. This statement is supported by data from the Global Islamic Economy Report (GIER). In 2019–2020, Indonesia's halal food consumption reached 173 billion US dollars (Yurivito 2019). In fact, this phenomenon can be an opportunity for Indonesia to become the largest halal industry producer in the world. This opportunity can be realized by maximizing the potential of the leading sectors of the halal industry in Indonesia, including halal food, halal tourism, Muslim fashion, integrated agricultural creative industries, and renewable energy (Sulaiman 2019).

The halal industry's huge potential can be seen from the development of several sectors in Indonesia, such as the food and beverage sector can contribute up to 3.3 billion USD from Indonesia's exports to Islamic Cooperation Organization (ICO) countries and non-OIC countries in 2017. The next sector that has a big potential is Muslim fashion. This can be seen from the Ministry of Industry data that about 30 percent of small, micro, and medium enterprise (MSMEs) industry players are engaged in Muslim fashion (Yolanda 2019). These sectors have great potential to be exported to boost the productivity of the halal industry in Indonesia.

## 4.3 Role of halal industrial estates in boosting national economic growth

The halal industry does have a strategic role in improving the national economy. This potential can already be seen from the data of the Ministry of Finance of the Republic of Indonesia in 2019 which states that the halal industry has contributed 3.8 billion USD to Gross Domestic Product (GDP) and 1 billion USD of investment from foreign investors, as well as opening 127 thousand jobs per year. In fact, if this potential is optimized again, the halal industry can increase the value of exports and foreign exchange reserves of the country. In addition, the halal industry also always shows a positive trend every year. This can be seen through the contribution of the Sharia economy to Gross Domestic Product (GDP) of 3.8 billion USD annually. Besides, the halal industry also has the potential value of Indonesia's exports, which is one of the driving factors of economic growth. The value of exports that can be produced from the halal industry ranges from 5.1 billion USD to 11 billion USD annually. In 2018, the halal industry generated 7.6 billion USD (Fathoni et al. 2020).

The halal industry's huge potential encourages the government to immediately create a new strategy to achieve these opportunities by establishing a halal industrial area. This area's concept is formed from existing industrial estates, but with adequate facilities and systems to produce halal products following the Halal Product Guarantee (HPG). The initiation of the establishment of the Halal Industrial Estate is a form of Indonesia's efforts in pursuing the progress of other countries in terms of the halal product industry (Yurivito 2019).

The Ministry of Industry, together with stakeholders, seeks to accelerate the growth of Sharia economy from regional aspects, namely encouraging the preparation of the halal industry ecosystem with the issuance of the Regulation of the Minister of Industry No. 17 of 2020 on Procedures for Obtaining Certificates in the Framework of The Establishment of Halal Industrial Estates. The regulation is a guide for industrial estate managers in improving infrastructure facilities and infrastructure supporting halal industry activities and guiding the halal industry in creating halal industry manufacturing, centralized and located in the halal industrial area.

Indonesia officially has the first Halal Industrial Area, the modern Halal Valley in Cikande, Banten. The Director of Sharia Economic Development and Halal Industry of the National Committee for Islamic Economics and Finance (KNEKS) said this was a concrete step to realize the great strategy of the Islamic Economic Development plan of Indonesia. The establishment of the Halal Industrial Estate must go through the conformity verification process established by the Ministry of Industry and the Halal Product Assurance Regulatory Agency (HPARA) and the Indonesian Ulama Council (MUI). In addition, it is also thanks to the encouragement given by KNEKS, which consists of the Ministry of Industry, Ministry of Religious Affairs, and other stakeholders.

The presence of this Halal Industrial Area will encourage the development of the halal products industry in Indonesia to be more advanced. Including producing more and more halal products of good quality for the domestic consumption market and to serve the needs of the world halal products. In addition, the development of the Halal Industrial Estate is also the first step to improve Indonesia's current position as a consumer country of halal products. As we advance, Indonesia can become the world's leading producer of halal products with an integrated halal area.

This shows that the halal industry market has contributed significantly to the development of the global economy. Nowadays, the use of halal products is motivated by religious demands or the large population of Muslims in the world and the growing awareness that halal products are quality products and guarantee consumer safety.

## 5 CONCLUSION

Indonesia has considerable opportunities and the potential for the development of the halal industry. If this potential is maximized, it will contribute to national economic growth. Optimization of the halal industry in Indonesia can be implemented through the development of Halal Industrial Estates. This region exists to create a new climate in the development of industrialization in Indonesia from capitalist to Sharia-based. This transition is a very appropriate step, considering Indonesia has an Islamic State background with the world's largest Muslim population. Indonesia is also ranked 10th in the world in terms of the application of the Sharia economy. These backgrounds have made Indonesia gain high confidence in developing the halal industry in the global halal market. Optimization of industrial estate development creates an increase both in national and global economy.

## REFERENCES

Alrasyid, M. H. 2016. Environmental Strategic Management for Green Industrial Area. *IJEEM – Indonesian Journal of Environmental Education and Management*, 1(1), 101–116.

Annisa, A. A. 2019. Kopontren and the Halal Value Chain Ecosystem. *Scientific Journal of Islamic Economics*, 5(01), 1. https://doi.org/10.29040/jiei.v5i01.398

Anwar Fathoni, M., Hadi Syahputri, T., Economics and Business, F., & National Development Veteran Jakarta, U. 2020. Portrait of Indonesia's Halal Industry: Opportunities and Challenges. Scientific Journal of Islamic Economics, 6(03), 428–435.

Elasrag, H. 2016. Halal Industry: Key Challenges and Opportunities. *SSRN Electronic Journal, 69631.* https://doi.org/10.2139/ssrn.2735417.

Dar, H., et al. 2013. *Global Islamic Finance Report: GIFR 2013*, Edbiz Consulting.

Global Leader. 2008. World Population Prospects the Islamic Bombs. Retrieved November 17th, 2009, from http://www.freeworldacademy.com/globalleader/population.htm.

Hakim, R. 2019. The urgency of halal Special Economic Zone (KEK). *Archives of Scientific Publications Bureau of Academic Administration.*

Istiqomah, N., & Prasetyani, D. 2013. Analysis of the impact of the existence of industrial estates in the village needs to increase economic activity of the community in mojosongo subdistrict boyolali district. *Journaling & Proceeding FEB Unsoed, 3*(1), 1–11.

Kamaruddin, R. and Jusoff, K. 2009. An ARDL Approach in Food and Beverages Industry Growth Process in Malaysia. *International Business Research, vol.2 no.3 pp.98–107.*

Ministry of Industry. 2017. Reached 5.49 Percent, Industrial Growth Skyrocketed Above the Economy.

Ministry of Finance of the Republic of Indonesia. 2019. "Financial Media." Ministry of Finance 14(140).

Majlis, F. H. 2011. The Market Opportunity of the Muslim World.

Muhyiddin, N. T., Tarmidzi, I.M., Yulianita, A. 2017. Economic & Social Research Methodology. Jakarta: Publisher Salemba Empat

Nisha, N., & Iqbal, M. 2017. Halal ecosystem: Prospect for growth in Bangladesh. *International Journal of Business and Society, 18*(S1), 205–222.

Othman, P., Sungkar, I. and Sabri Wan Hussin, W. 2009. Malaysia as an International Halal Food Hub: Competitiveness and Potential of Meat-based Industries. *Asean Economic Bulletin, vol.26 no.3, pp. 306–320.*

Othman, R., Ahmad, Z.A. and Zailani, S. 2009. The effect of institutional pressures in the Malaysian halal food industry. *International Business Management, vol.3, no.4, pp. 80–84.*

Sagala, Aryanto, et al. 2004. *Preparation of Industrial Estate Development Plan.* Jakarta: BPPT Press.

Sari, A. N. 2018. *Implementation of supervision of The Center for Drug and Food Control and Institute of Food, Medicine, and Cosmetics Assessment of Majelis Ulama Indonesia as an effort to protect consumers on cosmetic products in Serang City.*

Suheri, T., & Aulia, S. S. 2017. Triple Helix Analysis in a Special Economic Zone (Case Study: Sei Mangkei KEK). *Proceedings of the National Seminar on Engineering, Computer and Engineering,* v51–v58. http://prosiding-saintiks.ftik.unikom.ac.id/jurnal/analisis-triple-helix-dalam.46

Sugiyono, I'm sorry. 2005. Administrative Research Method. Bandung: Alfabeta Sulaiman, Dawn. 2019. Indonesia Has a Chance to Become a Halal Industry Producer. (https://www.wartaekonomi.co.id/read255374/indonesia-punya-peluang-jadi-produsen-industri-halal). Retrieved October 6, 2020.

Suryani, N. I., & Febriani, R. E. 2019. Special Economic Zone and Regional Economic Development: A Literature Study. *Convergence: The Journal of Economic Development, 1*(1), 40–54.

Syahruddin. (2010). Evaluation of Implementation of Industrial Estate Development Policy. *Journal of Administrative and Organizational Sciences, 17*(1), 31–42.

Tieman, M. 2007. Modern Cluster Applications for Age-Old Halal. *The Halal Journal, pp. 45–48.*

Yolanda, Friska. 2019. Great Potential of Indonesian Halal Industry.

Yurivito, Benedictus. 2019. Indonesia Is a Major Player in the World Halal Industry.

Yusof, R. 2011. Development of National Industry and The Role of Foreign Investment (PMA)-Rohalia. Yusof. *EKonomi And Education,* 71–87.

# Pesantren understanding on halal product guarantee

A.M. Nasih* & A. Sultoni
*Universitas Negeri Malang, Malang, Indonesia*

ABSTRACT: This study investigates *pesantren* people's understanding of halal products and halal product guarantee systems according to Law No. 33, of the Year 2014. These people's comprehension of halal products is essential since considering their substantial number, they held a considerable role in realizing the government-designated halal product guarantee system. This study used descriptive qualitative design classified as field research. The analysis results identify that *pesantren* people utilize the Islamic law (*Fiqh*) standards and understanding to ensure whether a product is halal. Those understandings and standards transform into the fundamental value that establishes the halal value chain in *pesantren*. Therefore, *pesantren* people own immense comprehension and awareness of the halal product. However, in relation to the halal product guarantee system determined by the government, *pesantren* still needs socialization and capacity development.

*Keywords*: pesantren, halal product, Islamic law

## 1 INTRODUCTION

The availability or clarity of information regarding the halal status of food, cosmetics, fashion, medicines, or other consumers' goods remains crucial for Muslims (Charity 2017). From the consumer's perspective, it involves faiths and religious doctrines (Ruslan et al. 2018), as well as human rights. Islam instructs only to have *halalan-thayyiban* (QS al-Baqarah [2]: 168). Foods designated so are halal, hygienic, and good for our health (Wibowo et al. 2020). Further, in the current modern era, the manufacturing industry has experienced significant development, along with the growth in science and technology. The increasing production in this sector has raised concerns about the quality of their products.

Consequently, as a form of consumer protection, the Indonesian government enacted Law No. 33 of the Year 2014 on a halal product guarantee, which is regulated in Government Regulation No. 31 of the Year 2019. The enactment of these regulations is important since it responds to the global demand on the halal industry (a global halal trend), and it shifts Indonesian halal certification from community-centric into state-centric (Akim et al. 2019). Previously, people or businessmen would indicate halal certification on their products to enhance their brand. Currently, that obligation has shifted onto the government. The government must ensure the quality of all products in Indonesia through halal certification. Besides, the government also ensures that each religious follower can worship and practice their religion safely through the halal guarantee for the products used and consumed by society.

The state-centric certification system relocation, as mentioned in paragraph 30 article 3 Government Regulation No. 31 of the Year 2019, creates an opportunity for the government to have corporation with society or religious institutions with legal certification. Thus, one of the influential entities that should be the government focus on implementing a halal product guarantee system is *pesantren*.

*Pesantren* is an inseparable institution from Muslims in Indonesia since Indonesia's growth and development are part of Islam's development (Shaleh 1982). Its presence as an Islamic religious

---

*Corresponding author

DOI 10.1201/9781003189282-14

institution has been proven strategic (Haningsih 2008) to shape clerics with in-depth spiritual knowledge, as well as awareness, skill, and concern of public issues (Haningsih 2008; Nasih 2014). In addition, *pesantren* is also productive in constructing Islamic regulations through a *fatwa* as an effort to solve the problems that appeared in society (Nasih 2014). *Pesantren* frequently refers to *turats* books in determining the regulations (Nasih 2014). Those regulations also include specifying the halal level of foods and beverages. This study aims to investigate *pesantren* people's comprehension of halal products and the halal product guarantee system regulated in Law NO. 33 of the Year 2014. Their comprehension is essential since they have a considerable number of resources that can help to implement the regulation.

## 2   METHODS

This research used qualitative method classified as field research and was carried out in middle of 2020. The research subjects are five *pesantren* in East Java categorized as modern (*pesantren* Attanwir), semi-modern (*pesantren* Bahrul Maghfiroh), and Salafi (*pesantren* Assuniyah and Lirboyo). The data were obtained through interviews, observation, and documentation. The obtained data was analyzed using content analysis suggested by Klaus Krippendorff (1993). The data analysis stages involve unitizing the data selected for research sources, sampling from various information and data sources, reducing, inferring, analyzing, and narrating. The analyzed data was presented and commented on in the research results and conclusion.

## 3   RESULTS AND DISCUSSION

### 3.1   *Pesantren as a living discourse*

Historically, *pesantren* has a continuously developing definition. The definition changes following society's needs. However, it has two primary functions as an educational and socio-religious institution. The changes are usually fundamental and practical. Formerly, people come to *pesantren* only to learn Islamic teaching and ways to recite the Qur'an. Currently, however, they also go to *pesantren* to get formal education because of job reason. Therefore, in the future, people will need *pesantren* of economic factor. This is because *pesantren*, as illustrated by Marzuki Wahid, is a living discourse. Therefore, discussion about *pesantren* is always fresh, alluring, practical, and difficult since it is multidimensional (Marzuki 1999). Marzuki explains that even with its multidimensional features, *pesantren* has the confidence and self-defense to face outside challenges (Marzuki 1999).

*Pesantren* is known as a traditional educational institution, but with its features and resources, it develops the most robust survival mechanisms in facing modern challenges. Thus, society often sees it as the best alternative to facing life challenges, including becoming the primary reference to conducting daily activities, primarily those related to Islamic principles and teachings. This is illustrated in perspective from Abdurrahman Wahid (Gus Dur) that cultural conservation is reflected in the *pesantren* intellectual tradition. The lessons offered in *pesantren* are in the form of universal literature through the guidance and blessings of *kyai* as the primary teacher (*irsyadu ustazin)*. The universal literature is in the form of yellow books, maintained and transmitted intergenerationally, and directly related to the unique concept of *kyai* leadership. The lessons contained ancient books (seen from a modern perspective) promise the right tradition (*al-qadimi ash-shalih)* and preserve the religious knowledge widely given to Muslims by past great clerics.

### 3.2   *Map of pesantren comprehension on halal product guarantee system*

*Pesantren* owns two primary functions as an educational and socio-religious institution. As an educational institution, it facilitates the teaching of religious knowledge such as the Qur'an, hadith, Arabic, *aqidah*, and *fiqh* (Fathoni 2005). Meanwhile, as a socio-religious institution, people who need the *pesantren* head into social and economic projections that become their fundamental needs (Marzuki 1999). Besides, *kyai* is an economically independent figure. Thus, the students (*santri*) in *pesantren* are also independent; in their tradition, they take care of themselves fully, from their consumption to their accommodation. In the fifth model, *pesantren, santri* are allowed to manage the kitchen, canteen, grocery store, and other related activities. In other words, without the parents, *santri* can live independently with their friends in *pesantren* (Zarkasyi 2015).

In the religious understanding context, such as deciding the halal and haram status of acts, foods, and beverages, *pesantren* possess an excellent understanding that is implemented in their daily life. This condition differentiates *pesantren* from other educational institutions in Indonesia. The other factors that differentiate them are that *pesantren* provide boarding houses (*pondok*) for the student's accommodation, their students are known as *santri,* the mosque used as a praying facility and a center for activities, *kyai* as a figure with religious competence and great charisma, and well as yellow books as the primary literature for Islamic studies (Syafe'i 2017).

Salafi *pesantren* or traditional *pesantren* are those who maintain learning using classic books with no general knowledge. The learning models commonly implemented in this *pesantren* isare *Sorogan, Wetonan,* and *Bandongan* (Anhari 2007; Dhofier 1985; Qomar 2005). This *pesantren* remains connected to the ancient *pesantren* or known as Salafi *pesantren.* This *pesantren* has several characteristics. First, its main activity is the literal study of classic books. Second, the method used in this *pesantren* is traditional methods, such as *sorogan, wetonan,* and memorization. Third, it does not use a structured class system in a study period, so that the learning achievement of *santri* is measured based on the number of books they have learned and with which cleric they learn it. Fourth, it the educational aim is to improve morals in acts and behaviors, as well as preparing the *santri* to implement the knowledge and morals in their daily life (Zarkasyi 2015).

Within the halal product guarantee system that has been implemented in *pesantren*, they use the fiqh standards that they have learned in the yellow books. Assunniyah Pesantren Kencong Jember is a Salafi *pesantren*. The institution was established in 1942 by K.H. Djauhari Zawawi in Kencong village, around 500 meters from the Kencong district office and 45 kilometers in the southwest of the center of Jember city. If a *pesantren* implements an educational system, then it uses a traditional (*madrosi*) system that consists of *ibtidaiyah (elementary), tsanawiyah (secondary),* and *'aliyah (secondary).* From those three education levels, the books remain are used in accordance with the *santri's* capability.

For the legal and Sharia (fiqh) learning, *pesantren* decides the books that should be learned, such as *Fathul Mu'in, Tahrir, Fathul Wahab,* as well as other *mu'tabaroh* books. Therefore, even in legal determination, those books are still used. For instance, in the determination of animals to be consumed, such as chicken, the committee of Assunniyah Pesantren Kencong Jember believe that the animal sold in the market around the *pesantren* are halal if the seller is Muslim, and it is also located in the Islamic area (*darul* Islam).

> For us, any doubt about an animal in an Islamic area (darul Islam) should be erased, since it should be halal, as explained in al-asylum fi alasyya' al hillu (the origin of things is halal) principle. However, if it is in an area with a non-Muslim majority, the doubt should be perceived as haram.
>
> (Interview by Musthofa 2020)

The same thing is also considered by the *santri* in evaluating food and clothing. The *pesantren* cooperate to fulfill the *santri* daily needs. The fulfillment of some requirements also involves the surrounding society. The *santri* also understand which product is halal or haram in selecting foods or beverages. Besides, each product provided in the cooperative is selected by the *pesantren* management. For example, a perfume with alcohol above the standard is prohibited from being traded in the cooperative. Meanwhile, for the standards of the halal product guarantee enacted by the government, the *pesantren* have not understood it since they have not received socialization.

The same finding was also found at other *Salafi* pesantren, Lirboyo Pesantren Kediri, East Java. Lirboyo is the name of a village used as the *pesantren's* name. The history of this *pesantren* cannot be separated from its founder KH. Abdul Karim, who moved and stayed in this village since 1910. Other than being the center of Islamic education, Lirboyo is also well-known as Salafi *pesantren* that is successful in combining tradition that complete the modernity and has brought many religious-pious and socially pious figures.

According to the statement of one senior cleric (Interview by Ali Muntaqo 2020), even if it is as famous as Salafi *pesantren,* Lirboyo possesses various educational systems. The system used depends on the policy of the tutor and management in every unit. However, even if the implemented methods are different, all units must comply with the Welfare Development Body of the Lirboyo Pesantren. This body has the highest authority in the Lirboyo Pesantren. Every activity or policy in the boarding house or school can only be realized if it is approved by that body. If there is an unresolved issue in the central or unit of the *pesantren,* then it is taken to the Welfare Development Body of the Lirboyo Pesantren meeting.

Other than that body, the educational system is also managed by the Deliberative Council of Lirboyo Pesantren that is now called the Lajnah Bahtsul Masail of Lirboyo Pesantren. This council also produces legal fatwas. This council has the most considerable strategy role for three main reasons: first, it acts as a mediator in socializing new ideas of Islamic understanding to society. Second, it functions as a body that sharpens the skills, creativity, and intellectual quality of the *santri* in various religious sciences, primarily *fiqh.* Third, it prepares competent *santri* in accommodating different thought in society and formulate a sensible and wise decision.

In the halal product guarantee, Lirboyo has completed studies on some food, fashion, and medical products. As explained by Ali Muntaqo (2020), Lajnah Bahtsul Masail of Lirboyo Pesantren always publishes books from the legal discussion (*bahtsul masail*) in each educational level. In other words, the body has unique authority in deciding and regulating policy, including the halal status of a product. Further, after the regulation or new ideas on Islamic teaching has been enacted, it is socialized to the society through guideline books or directly delivered in da'wah. Interestingly, this body has also published a book entitled, Medical *Fiqh*: Study of Sharia Law, History and Lesson in Medical World. Like the Assuniyah Pesantren Jember, Lirboyo Pesantren also has not technically understood the standards of the halal product guarantee regulation. However, they have decided the standards for halal product guarantee to determine the halal status of a product.

*Semi-modern* pesantren is *pesantren* combining Salafi and Khalaf *pesantren.* The *pesantren* conduct its education with a hybrid system. It uses traditional book learning with formal *madrasa* and adopts a government curriculum (Alwi 2013). In other words, a semi-modern *pesantren* integrates traditional and modern *pesantren.* Other than the traditional *pesantren* curriculum, the classic book learning in this *pesantren* also uses the Ministry of Religious Affairs curriculum and the Ministry of Education curriculum (Syafe'i 2017). This kind of *pesantren* keeps its *Salafi* part by the study of classics books, and the *Khalaf* aspect of this pesantren is reflected from its formal education such as *madrasa,* schools, and universities. The study of classical books is obligatory for all of its santri in all levels of education, even if some of the *santri* only join the classical books study. This *pesantren* also provides non-formal Islamic education (*Majelis Taklim)* and vocational education (Zarkasyi 2015).

In this study, Bahrul Maghfiron Pesantren, Malang was selected as representation of a *semi-modern pesantren.* This *pesantren* was pioneered in 1995; however, it was built in 1997, simultaneously with the inauguration of Mbah Kiai Fattah Mosque, as the founding father of the *pesantren.* He later appointed his son, Lukman Al Karim bin Abdullah Fattah as the head of the tutor. This *pesantren* develops several methods, including traditional methods, such as *sorogan* and *wetonan,* as well as the combination method by learning various *mu'tabarah* books that are frequently used by *Ahlussunna wal jamaah* clerics.

In addition, Bahrul Maghfiron Pesantren also develops economic sectors through businesses. The developed businesses use standards explained in yellow books. In other words, they understand the products to be developed in accordance with the Islamic Sharia. Therefore, the established business units mostly used Sharia labels, such as BM (*Bahrul Maghfiroh*) Kitchen as a business unit of the Sharia Cooperative owned by Bahrul Maghfiroh Al-Fattah As-Syafi'iyah Foundation. Other than to improve the *pesantren* economy, BM Kitchen also focusses on home cooking that is pleasant, inexpensive, and *halalan thoyyiban* based on Islamic Sharia. BM Kitchen started to operate on October 9, 2018 and was inaugurated by East Java Governor Hj. Khofifah Indar Parawansa on November 9, 2018. In addition to BM Kitchen, they also have Bahrul Maghfiroh Mart (BM Mart), a grocery store that aims to realize an independent economy that aids the *pesantren* operational cost. This store has not been affiliated with halal companies, but its products are halal products produced by small and medium enterprises.

Additionally, it also develops BM Urban Farming in the agricultural sector. By using the hydroponics method, it utilizes a 12x8 meters area as agrarian land with 5000 holes. The plants produced by them include red spinach, green spinach, red lettuce, kale, celery, and so forth. The superiority of hydroponic vegetables is because they are healthier, pesticide-free and more nutritious. Besides, the hydroponics method also preserves the halal level of the products.

Modern *pesantren,* or known as Khalaf *pesantren,* also implement classic learning (*madrasa),* provide general and religious sciences, as well as provide vocational education. This *pesantren* has undergone modernization in its educational system, institutional system, speculation, and function, along with technological support and proper foreign language institution (Zarkasyi 1998). In relation to this, *ma'had aly* is also categorized as modern *pesantren.*

In addition, the modern *pesantren* provides a library, a public kitchen, a living room, a dining room, an administrative office, a grocery store or cooperative, a meeting hall, a toilet, and adequate laboratory. The activity in modern *pesantren* involves book study, madrasa, and public elementary school to university activity, as well as vocational education and environmental development education (Saridjo 1979).

Attanwir Talun Pesantren, Bojonegoro, was selected as representation of modern or *kholaf pesantren.* It was established in 1993 by KH. M. Sholeh. Formerly, it only taught the children to fluently recite the Qur'an, writing in Arabic, and prayer procedures, which are currently only found in *pesantren* with the *wetonan* system. In 1951, this *pesantren* started to use the classical learning system by creating *diniyah* classes with a two-year learning period. Later, in 1954, its religious educational system developed into *Madrasa Ibtidaiyah* with 6 years of learning. Then, to accommodate the *Madrasa Ibtidaiyah* graduates, it established Madrasah Mu'allimin Al-Islamiyah with a 4-year study period in 1961. Following the advancement and demand that are continuously growing, it was changed into Madrasa Tsanawiyah Islamiyah for 3 years and Madrasa Aliyah Islamiyah for 3 years, as well all Sekolah Tinggi Agama Islam (STAI) Attanwir. All the Islamic learning processes, including fiqh, in those levels, are perceived to be sufficient to guide the *santri* to live in a society (Interview by Sahal 2020).

In relation to the halal product guarantee, Attanwir Pesantren has implemented a system a long time ago. This system follows the fiqh that has been taught to the *santri.* Besides, Attanwir Pesantren has also implemented Halal Mart that provides foods and medical products for the *santri.*

> *Pesantren has credible religious integrity, including the products produced by the pesantren. For instance, Attanwir had always produced halal beverages, however, the production cannot be continued. The santri consumption is managed by the pesantren's canteen based on the standards we have*
> (Interview by Sahal 2020).

Sahal (2020) also mentioned that Attanwir always maintains the quality of food consumed by the *santri,* including the purchase of the consumed animal. Every purchase of meat, even if it is rare, is always done with a *pesantren* graduate Muslim seller who knows how to slaughter animals based on Sharia (Interview by Sahal 2020). Consequently, the *pesantren* understands the halal level of *santris'* needs, as well as society's needs of the *pesantren* integrity as a religious, educational and Islamic institution.

Additionally, Attanwir Pesantren has also opened the Halal Network International (HNI) business center store which is affiliated with Herba Penawar Alwahida Indonesia (HPAI). It is managed by an administrator in the Rumah Sehat LKSM Attanwir office (Interview by Muflihatin 2020). Further, this *pesantren* has also developed waste management to empower *santris'* economy by producing fertilizer, crafts, and so forth. The organic waste was managed into fertilizer. While, for the crafts, the inorganic waste, such as paper, plastics, and so forth, were used. In other words, the waste went through a feasibility test before the production process. For the regulation related to the halal product guarantee enacted by the government, Attanwir has even received an invitation to send its delegate to join the Halal Product Guarantee training in a university in Surabaya. However, due to the time and condition constraints, the *pesantren* did not send its representative (Interview by Sahal 2020).

## 4 CONCLUSION

The Salaf, semi-modern, and modern *pesantren* have properly comprehended the halal and haram standards for food and beverages products. As mentioned in Islam, all foods and beverages are halal, except those declared harams, such as corpses, blood, pork, and animals slaughtered not in the name of Allah (QS Al-Baqarah [2]: 173). Meanwhile, haram beverages include all forms of drinks that contain alcohol (QS. Al Baqarah [2]: 219). This idea has been adequately understood by the *pesantren* people following their studies on relevant classic books. The study of classic book in the *pesantren* has become fundamental aspect in ensuring that every good consumed or used is halal. It is proven that those understandings become the base in establishing halal chain value in *pesantren*. In other words, *pesantren* has excellent awareness of halal products. However, in relation to the halal product guarantee regulation enacted by the government, further socialization is still required to develop the capacity.

## REFERENCES

Akim, N. K.; Chandra, P. & Leej, C.K. 2019. The Shifting of Halal Certification System in Indonesia: From Society-Centric to State-Centric, *MIMBAR*, Vol. 35 No. 1st: 115–126.

Alwi, B.M. 2013. Pondok Pesantren: Ciri Khas, Perkembangan, dan Sistem Pendidikannya. *Lentera Pendidikan* Vol. 16, No. 2: 205–219.

Anhari, M. 2007. *Integrasi Sekolah ke dalam Sistem Pendidikan Pesantren.* Surabaya: Diantama.

Charity, M.L. 2017. Jaminan Produk Halal di Indonesia. *Jurnal Legislasi Indonesia* Vol. 14 No. 01 — Maret: 99–108.

Dhofier, Z. 1985. *Tradisi Pesantren: Studi Tentang Pandangan Hidup Kyai.* Jakarta: LP3ES.

Fathoni, M.K. 2005. *Pendidikan Islam dan Pendidikan Nasional: Paradigma Baru.* Jakarta: Direktorat Jenderal Kelembagaan Agama Islam.

Haningsih, S. 2008. Peran Strategis Pesantren, Madrasah dan Sekolah Islam di Indonesia. *El Tarbawi: Jurnal Pendidikan Islam* No. 1. Vol. I: 27–39.

Sahal, M. Rafiq, *Interview* in Pondok Pesantren Attanwir Talun Bojonegoro. 28 June 2020. Krippendorff, K. 1993. *Analisis Isi Pengantar Teori dan Metodologi.* Jakarta: Raja Grafindo.

Marzuki, W., (etc.) 1999. *Pesantren Masa Depan: Wacana Pemberdayaan dan Transformasi Pesantren.* Jakarta: Pustaka Hidayah. Muflihatin, Siti Imaniatul, *Interview* in home of alumni Pondok Pesantren Attanwir Talun Bonojegoro, 29 June 2020.

Muntaqo, Ali, *Interview* in Pondok Pesantren Lirboyo Kediri. 5 July 2020.

Musthofa, *Interview* in Pondok Pesantren As-Suniyah Kencong Jember. Sunday, 21 June 2020.

Nasih, A.M. 2014. Pemaknaan masyarakat santri atas fatwafatwa keagamaan produk pesantren. *Ijtihad: Jurnal Wacana Hukum Islam dan Kemanusiaan* Volume 14, No. 1, Juni: 119–138.

Qomar, M. 2005. *Pesantren dari Transformasi Metodologi Menuju Demokratisasi Institusi,* Jakarta: Erlangga.

Ruslan, A. A. A., Kamarulzaman, N. H. & Sanny 2018. Muslim consumers' awareness and perception of Halal food fraud. *International Food Research Journal* 25 (Suppl.1), December: S87–S96.

Saridjo, M. 1979. *Sejarah Pondok Pesantren di Indonesia.* Jakarta: Dharma Bhakti.

Shaleh, A.R. (etc.) 1982. *Pedoman Pembinaan Pondok Pesantren* Jakarta: Proyek Pembinaan dan Bantuan Pondok Pesantren.

Syafe'i, I. 2017. Pondok Pesantren: Lembaga Pendidikan Pembentukan Karakter. *Al-Tadzkiyyah: Jurnal Pendidikan Islam* 8 (1): 61–82.

Wibowo, M.M., Hanafiah, A., Ahmad, F.S., Khairuzzaman, W. 2019. Introducing Halal Food Knowledge to the Non-Muslim Consumers in Malaysia (Its Effect on Attitude and Purchase Intention). *Advances in Economics, Business and Management Research* Volume 120, 4th International Conference on Management, Economics and Business (ICMEB).

Zarkasyi, A.S. 1998. *Langkah Pengembangan Pesantren dalam Rekontruksi Pendidikan dan Tradisi Pesantren Religiusitas Iptek.* Yogyakarta: Pustaka Pelajar.

Zarkasyi, H.F. 2015. Sistim Pendidikan dan Pengkajian Islam di Pesantren dalam Kontek Dinamika Studi Islam Internasional. *EDUKASI: Jurnal Peneliti an Pendidikan Agama dan Keagamaan,* Volume 13, Nomor 3, Desember: 335–348.

*Halal Development: Trends, Opportunities and Challenges – Pratikto et al (eds)*
© 2021 The Author(s), ISBN 978-1-032-03830-8

# Islamic work ethics in management performance perspectives: Conceptualization and value internalization

J. Hutasuhut, A.R. Syamsuri & A. Saragih
*Universitas Muslim Nusantara Al Washliyah Medan, Indonesia*

S. Sarkum*
*Universitas Labuhan Batu, Rantauprapat, Indonesia*

ABSTRACT: Islamic work ethics can be understood as a set of values or belief systems based on the Qur'an and the Hadith regarding work and hard work. The purpose of Islamic work ethics is to align the organizational goals with employees' goals so that they can run the business well. With such attitudes and behaviors, it is expected that companies or organizations be able to adapt and respond to various changes in a business environment that are changing rapidly. The benefits of Islamic work ethics are expected to be a requirement for employees to work so they will be safe in both the world and hereafter according to the Islamic values which are determined in the Qur'an and Al-Hadith. Workers with the Islamic work ethic will be more motivated to produce optimal performance hope that they will get halal sustenance and will avoid disapproval from Allah SWT.

*Keywords*: Islamic work ethics, halal, management, performance

## 1 INTRODUCTION

Etymologically, ethics and morals both mean customs. Ethics comes from the ancient Greek 'ethos' and plural '*taetha*', while morals comes from the Greek '*mos*' and the plural 'mores' (Bertens 2007). Ethics deals with questions about how people will behave towards each other (Hidayah 2018). Ethics as a reflection is moral thinking which means that humans must think about what must be done or not to be done. Philosophically, ethics has a broad meaning as a study of morality. It also encompasses several functions. First, descriptive ethics describes moral experience to determine the motivation and purpose of an action in human behavior. Second, normative ethics attempts to explain why humans act as they do and what principles emerge in humans' lives. Third, metaethics attempts to give meaning, terms, and language for use in ethical discussions and give the ways of thinking used to justify ethical agreements (Nawatmi 2010). The organization cannot answer many human resource problems through rationality, such as by local intellectuals to western management. This phenomenon is seen in the results of studies that continue to roll from progressive Islamic thinkers in countries with a Muslim majority, such as Indonesia. The individualist, materialist, secular nature of western management and manipulative practices have reinforced the argument that Muslims must not forget the religious values regulated by Allah and the Messenger of Allah. Islam as a religion that brings the concept of *rahmatanlilalamin,* and provides guidelines for people, including how their work organizes ethics in an organization (Kusuma & Ralmat 2018).

In relation to ethics, Islam suggests that every individual who works will tell him/her about what he/she has done. Islam also explains halal and haram. Ridwan (2019) explains that, in Islamic Sharia, halal is the opposite of haram. Conducting haram will incur punishment and doing halal

---

*Corresponding author

DOI 10.1201/9781003189282-15

will be rewarded. Allah is all-knowing and all-compassionate (Kurniawan 2019). About the rules of working in Islam, mankind is commanded by God to always do beneficial work to oneself and others (Ali and Al-Owaihan). According to this explanation, the quality of human resources is not determined only from the knowledge of science and technology but also from the spiritual dimension of faith and taqwa through the dimension of the Islamic Work Ethic accompanied by a halal understanding or through the ordered teachings of Islam. Furthermore, by knowing the dimensions of the work ethic according to halal principles, an employee will have more spiritual responsibility towards all the resources of the organization used to achieve optimal performance. Effective management performance is one way to conduct an activity to be better and develop. Therefore, to practice management performance more effectively, normative rules are needed, containing a system of values and moral principles that are guidelines for employees in conducting their work duties. Aggregation of employee ethical behavior becomes a work ethic description to be applied. Islamic work ethics can be a part that can affect an employee's involvement in doing their work. According to Ali and Al-Owaihan (2008), Islamic work ethic is an orientation that shapes and influences the involvement and participation of its adherents in the workplace. Islamic work ethics enhance self-interest economically, socially, and psychologically, maintain social prestige, advance society's welfare, and reaffirm faith. The initial concept is derived from the al-Qur'an and the Sunnah.

In modern organizations nowadays, the Islamic work ethic has become a trend to be studied in-depth terms. The main factor is the attractions because there are many studies that prove that Islamic work ethics have a significant influence on employee performance. With this assumption, Islamic work ethics has potential to be used as a solution to mitigating performance problems. Some studies that support this assumption include research conducted by Hutasuhut (2019). He explored the influence of Islamic work ethics on the performance of lecturers at the Muslim Nusantara University of Al-Washliyah Medan. Participants involved were 47 tenured lecturers at the university. The results showed that an Islamic work ethics had a positive effect on working performance. It also states that Islamic work ethics have a close relationship with lecturer performance (Hutasuhut 2019). Lecturers as professional educators cannot just have skills, knowledge, and abilities but must have an Islamic work ethic to produce a more meaningful performance to the required standards.

Imam et al. (2013) analyzed features related to the employees' workings in the corporate sector. In this research, Siddiqui et al. (2019) consider religious values and human life in society and other aspects of life. Islam provides the rules of human life and the basic rules of human life. This study showed two main constructions of Islamic ethics: Islamic recruitment and selection, the Islamic compensation system and employee performance were used as dependent variables. Researchers used 100 employees from the Islamic Bank in Karachi, Pakistan. According to Siddiqui et al. (2019), although many studies focus on ethics in the workplace, ethics still becomes the most discussed matter in recent years. This is due to the continuation of unethical cases in organizations such as corruption, fraud, sexual harassment, and other unethical behavior. Abdi et al. (2014) used a random sampling method. Data was collected using a questionnaire distributed to 40 staff and lecturers at Insaniah University, Kedah. The first part of the questionnaire covered the demographic profile of the respondents, while the second part examined respondents' level of agreement on questions of Islamic work ethics, job performance, and organizational commitment items. The results show that job performance was significantly related to Islamic work ethics. On the other hand, the organizational commitment was not significantly related to Islamic work ethics.

The study conducted by Abbasi et al. (2012) investigated the influence of Islamic work ethics toward organizational learning, innovation, and performance. This research was conducted in the service industry in Pakistan to provide insight into how far the application of Islamic work ethics in business organizations is applied in Pakistan, especially in manufacturing. In this study, data was collected from 240 industrial service employees. The selected districts for data collection are Gujrat, Gujranwala, Sialkot, Lahore and Wazirabad. The main factory is in Gujrat, the head office in Lahore. The results support the opinion that Islamic work ethics in business organizations increase learning, innovation, and organizational performance.

Research conducted by Rohman et al. (2010) that explored the influence of the Islamic work ethic in work such as job satisfaction, organizational commitment and intention using a sample of 49 employees from 10 Islamic microfinance institutions in Demak Regency, Central Java, Indonesia, found that the Islamic work ethic has a positive effect on both job satisfaction and organizational commitment. Ibrahim (2018) conducted a study on the effect of Islamic work ethics on Muslim employees' efficiency in multinational companies. Research observations analyzed the potential and actual impact of Islamic work ethics in multinational corporations' environments by showing that religion does not significantly influence efficiency, but its internal effects such as internal and interpersonal conflicts with types of corporate activities can be intensified. In fact, this effect will differ based on the division of each employee. For example, when Muslims' religious orientation increases, their activities, according to the type of product or service provided, in the marketing sector can be affected.

Adisi (2014) explored the effect of job satisfaction in accountant performance and the impact of interactions between Islamic work ethics and job satisfaction to maximize accountant performance in the Islamic financial industry in Indonesia. The research instrument used to collect primary data was a questionnaire. Regression analysis was used to test the hypotheses. His study revealed that job satisfaction cannot affect the performance of accountants. He also unveiled that the interaction between Islamic work ethics with job satisfaction can maximize the performance of accountants. Amilin's (2016) efforts to maintain workers' level of performance excellence have been widely discussed, but practitioners still face difficulties, especially in Muslim countries. This is due to complex human behavior and many other factors, such as work stress and dissatisfaction. Although there is a lot of research about ethics has been done, both in an Islamic perspective or a general perspective, there is a lack of studies that investigate the impact of Islamic religious awareness in the workplace. Thus, in response to the impact of Islamic religiosity and Islamic work ethics in improving and maintaining performance, this research is presented, discussed, and investigated empirically for possible implications between Islamic religiosity, Islamic work ethics and performance.

Zahraha et al. (2016) discussed the issue of the work ethic with the rationale of the growing popularity of religion when Max Weber developed the theory of Protestant Work Ethics (PWE), which was then able to strengthen the Western economy through its ideology of capitalism. In his work, he claims that only Protestant Christians can accelerate and improve the economy, while other religions cannot do the same acceleration. Although many scholars reject the claim, the backwardness of many Islamic countries, ethically and economically, has raised many questions about Islam's teaching about work activities. Therefore, this study explores the concept of work ethic from an Islamic perspective, specifically annulling the concepts, principles, dimensions, and values of Islamic Work Ethics (IWE) to be operationalized into daily practice. This will also reveal the truth of the verses of the Qur'an and Hadith, which discuss the concept of work ethics in Islam. If properly operationalized, this can bridge the gap between IWE theory and the reality of economic development in Islamic countries. Obedience will affect economic development in Islamic countries, bringing happy human life in this world and in the hereafter.

## 2 ISLAMIC WORK ETHICS

Irham (2012) explains that work ethics or work ethos in Islam are the values contained in al-Qur'an and al-Sunnah about 'work', which are used as a source of inspiration and motivation by every Muslim to do their work activities in various fields of life. The way they understand, live and practice the values of the al-Qur'an and al-Sunnah about the motivation to work is what shapes the work ethic of Islam. Sono et al. (2019) argues that the Islamic work ethos or Islamic work ethic is a personal attitude that creates a very deep conviction that work is not only to glorify himself and reveal his humanity, but also as a manifestation of pious deeds and therefore has a very noble worship value. Islam is a *syumuliyah* religion (perfect in terms of teachings) in the sense of regulating the entire joints of human life, including in working as part of the caliph on earth. This is explained implicitly in the Qur'an. Amilin (2016) also discussed are the Islamic work ethics which are based on two dimensions, namely the *ukhrawi* and worldly dimensions. In the *ukhwari* dimension, Syari'ah emphasizes the importance of intention, which is solely to get virtue from

God. Work based on Syari'ah principles not only show the nature of Muslims but also elevates his dignity as a trustworthy servant of Allah. Theoretically, syari'ah teaches the concept of *ihsan*, which means doing perfect work, and *itqon*, which means a very earnest, accurate, and perfect learning process.

## 3  PARAMETERS OF ISLAMIC WORK ETHICS

Ghozali (2018) explains that the construct of Islamic work ethics was built from several indicators, such as good activity, usefulness, working towards perfection, working better, working together, optimal effort, competing, helping time, being honest, trustworthy, smart, and *tabligh*. Islamic Work emphasizes intentions over the result. Jufrizen and Parlindungan (2016) and Ali and Owaihan (2008) contend that Islamic work ethics parameters are pursuing legitimate business, earned wealth, quality of work, wages, reliance on self, monopoly, bribery, deeds and intention, transparency, greed, and generosity.

## 4  FUNCTION AND PURPOSE OF ISLAMIC WORK ETHICS

Islamic work ethics are a motivating factor for a person in doing his job well because the work is part of worship. The motivation that comes from within individuals, not from external rewards, is called intrinsic motivation. Islamic work ethics gives meaning to the work we do, that work in the world is one of the determinants of happiness in the hereafter. With these beliefs and awareness, a person's motivation will be greater for doing work even without any incentives (Amaliah et al., 2013). Some jobs will get different treatment depending on how a person sees and believes in his work. Allah Almighty has ordered His people that all their works will be seen by Allah, His Messenger, and those who believe. Someone who has a high work ethic will certainly not feel comfortable if they are not working for several days. In Islam, faith demands real proof through charity, and charity is good deed work. Then, a true Muslim's work ethic will bring up the *muhajadah* that is serious in realizing his ideals. As a set of positive behaviors at work that comes from beliefs, work ethics can be a driving force.

   Hidayat (2015) also corroborates this assertion, as explained in surah al-Dzariyat: 51. The purpose is normatively derived from the Islamic faith and the true mission of life that man and other beings are created for is "to worship the creator that is Allah SWT." That goal is essentially transcendental because it is limited to the individual's worldly life and life after this world. In achieving these goals, ethical rules are needed to ensure that efforts that realize both the people's objectives and operative goals are always on the right path.

## 5  WORK MANAGEMENT

Aprida (2018) stated that the concept of performance management is an idea that employees must perform their duties or work in accordance with established standards. According to halal philosophy, this concept is also explained in Islam. Good and quality workers will strive to increase their achievement by giving full commitment to the work done. Gruman (2011) also stated that quality workers would also set objectives that they believe will positively impact the organization in a vibrant work atmosphere.

   Performance management is a term widely used in human resources and has a special meaning related to the review and management of individual performance. Human resource management plays an essential role in helping organizations obtain and maintain competitiveness compared to high-performance work systems (Idris 2019). Ochurub et al. (2012) suggest that performance management is the ongoing process of identifying, measuring, and developing individual and team performance and aligning their performance with the strategic objectives of the organization.

Syamsuri et al. (2019) stated that performance management is a comprehensive process related to performance. This reflects the entity's approach to performance and includes subprocesses such as strategy definition, strategy implementation, training, and performance measurement.

## 6 CONCLUSION

Islam regards humans as creatures created by Allah to carry out the mandate of doing worship for their provisions both in this world and in the hereafter. In the context of performance management between the company and employees as leaders and employees of the Islamic kin gdom, human beings must have a high work ethic in carrying out all the mandates they receive to achieve the pleasure of Allah. Islamic work ethic is the work behavior of an employee in accordance with what is ordered and forbidden by Allah SWT as written in the Qur'an and Hadith. "Work for you, so Allah and his messenger and the believers see your work as if you were going to live forever." Islamic work ethics can also be understood as a set of values or belief systems based on the Qur'an and Hadith regarding work and hard work. Islamic work ethics can have a good impact on employee behavior at work because it can provide a stimulus for positive work attitudes. This matter becomes the principle that a Muslim's aim in working is to seek the pleasure of Allah SWT and get the virtue of quality and wisdom from the results obtained.

## REFERENCES

Abbasi A.S., Mir G.M. & Hussain M. 2012. Islamic Work Ethics:How It Affects Organizational Learning, Innovation & Performance. Actual Problems of Economics, 12, 471–478.

Abdi M. F., Nor S.F.D.W.M. & Radz. Md.N.Z. 2014. The Impact of Islamic Work Ethics on Job Performance and Organizational Commitment, Proceedings of 5th Asia-Pacific Business Research Conference 17–18 February, Hotel Istana, Kuala Lumpur, Malaysia, ISBN: 978-1-922069-44-3.

Adisi M. 2014. The Effect of Islamic Work Ethics on The Performance of Muslim Employees of Marketing Sector In Multinational Companies, 3, 31–40, International Journal of Organizational Leadership.

Afrida. 2018. Hakikat Manusia dalam Perspektif Al-Qur'an. 16(2), 54–59, Jurnal AL-QISTHU: Kajian Ilmu-Ilmu Hukum, IAIN Kerinci.

Amaliah I., Julia A. & Riani W. 2013. Pengaruh Nilai Islam terhadap Kinerja, 29(2), 165–174, MIMBAR.

Amilin A. 2016. Measuring the Correlation of Job Satisfaction with Accountants Performance: The Role of Islamic Work Ethics as a Moderator, European Research Studies Volume XIX, Issue 4, 217–232.

Ali A. J., & Al-Owaihan A. 2008. Islamic work ethic: a critical review", Cross Cultural Management: An International Journal, Vol. 15, No. 1, pp. 5–19.

Bertens K. 2007. Etika. Gramedia Pustaka Utama, Jakarta.

Ghozali M. 2018. Konsep Profesionalisme terhadap Pekerjaan dalam Perspektif Islam, Proceedings of the 5th International Conference on Management and Muamalah.

Gruman J. A. & Saks A. M. 2011. Performance management and employee engagement. Human Resource Management Review, 21(2), 123–136. 2011,

Hidayah N. 2018. Analisis Etika Kerja Islam dan Etika Penggunaan Komputer terhadap Ketidaksesuaian Penggunaan Komputer oleh Pengguna Teknologi Informasi di UMKM Kabupaten Bantul, 8(1), 59–73, Jurnal EkonomiSyariah Indonesia.

Hidayat S. & Tjahjono H.K. 2015. Peran EtikaKerja Islam dalam Mempengaruhi Motivasi Intrinsik, Kepuasan Kerja dan dampaknya terhadap Komitmen Organisasional (Studi Empiris pada Pondok Pesantren Modern di Banten). Jurnal Akuntansi dan Manajemen, Universitas PGRI Yogyakarta, 625–637, upy.ac.id

Hutasuhut J. 2019. Pengaruh Etika Kerja Islami terhadap Kinerja Dosen, 2(2), 1–3, Jurnal BEST: Biology Education, Scientific & Technology.

Ibrahim A. 2018. Islamic Work Ethics and Economic Development in Islamic Countries: Bridging Between Theory and Reality, 2, 43–50, Proceeding of 2nd International Conference on Empowering Moslem Society in Digital Era, ISSN 2622–5840.

Imam A., Abbasi A.S, & Muneer S. 2013. The Impact of Islamic Work Ethics on Employee Performance: Testing Two Models of Personality X and Personality Y. 25(3), 611–617, Sci.Int (Lahore).

Idris M., Zakaria W.F.A.W., Long A.S. & Salleh N. 2019. Kualiti Kerja dalam Organisasi: Tinjauan dari Perspektif Pengurusan Islam Work Quality in Organization: An Islamic Management Perspective, 15, International Journal of Islamic Thought.

Irham M. 2012. Etos Kerja dalam Perspektif Islam. 14(1), 11–24, Jurnal: Substantia. Jufrizen & Parlindungan R. 2016. Model Pengembangan Etika Kerja Berbasis Islam Pada.

Perguruan Tinggi Islam Swasta di Kota Medan. I (1), 120–136. Jurnal Ilmiah Maksitek.

Kusuma B. M. A. & Rahmat I. 2018. Mozaik Islam dan Manajemen Kinerja, Yogyakarta, Samudra Biru.

Kurniawan R. 2019. Urgensi Bekerja dalam Al-Qur'an, 3(1), 42–67, Jurnal Transformatif: Islamic Studies, IAIN Palangkaraya.

Nawatmi S. 2010. Etika Bisnis dalam Perspektif Islam, 9(1), 50–58, Fokus Ekonomi (FE).

Ochurub M., Bussin M. & Goosen X. 2012. Organisational readiness for introducing a performance management system. SA Journal of Human Resource Management, 10(1), 1–11.

Ridwan M. 2019. Nilai Filosofi Halal dalam Ekonomi Syariah. 3(1), 14–29, Profit:Jurnal Kajian Ekonomi dan Perbankan, STAIN Kudus.

Rokhman W. 2010. The effect of Islamic Work Ethics on Work Outcomes. EJBO Electronic Journal of Business Ethics and Organization Studies. 15(1), 21–27.

Siddiqui N.U.H., Hameed S. Sattar R. A. & Eneizan B. 2019. Islamic Work Ethics Impact on Employee Performance, 3(8), 1–7, International Journal of Academic Multidisciplinary Research (IJAMR).

Sono N. H., Hakim L. & Oktaviani L. 2019. Etos Kerja Islam sebagai upaya meningkatkan Kinerja. 411–420, e-Proceeding. https://jurnal.unej.ac.id/index.php/prosiding/article/view/6687

Syamsuri A.R., Halim A. & Sarkum S. 2019. Organizational Transformation:A Reviews of The Literature. Int. Journal of Scientific & Technology Research, 8(8).

Zahraha N., Norasyikin S., Rani S.H. & Rani B. A. 2016. The Relationship between Islamic Religiosity, Islamic Work Ethics and Job Performance, 710–716, The European Proceedings of Social & Behavioural Sciences Ep SBS.

# Halal responsibilities through Islamic business ethics practices: Implementation of trustworthy and fair values in traditional markets

H. Muhammad* & N.P. Sari
*Universitas Islam Raden Rahmat, Malang, Indonesia*

ABSTRACT:   This study was designed to examine the implementation of Islamic business ethics as a form of halal responsibility in business practices in the Tirtoyudo traditional market, Malang, Indonesia. An exploratory descriptive design with a single case study was used in this study. Data was collected through in-depth interviews with several participants who were selected purposively and with snowball sampling. The data was then analyzed using interactive models. The results showed that trustworthy and fair values underlie business practices as a halal responsibility in traditional markets. The successful application of trustworthy and fair values through a continuous coaching process is supported by market entities with existing local cultural synergies. The application of trustworthy and fair values in a mix of local cultures has implications for the sustainability of traditional markets which can be adopted by both traditional and modern markets as a halal market.

*Keywords*:   halal responsibility, Islamic business ethics, trustworthy, fair, traditional market

## 1   INTRODUCTION

Islam is a source of business ethics that leads to economic growth and development that has a positive impact on people's lives (Hassan 2016). Understanding business ethics with a strong religious foundation and being a philosophy of life has served as a catalyst for halal business transactions (El-Bassiouny 2014). This orientation is carried out by paying attention to correct business transactions as a form of halal responsibility for both products and businesses (Muhammad 2020). As a Muslim, the true halal responsibility is not only with fellow business partners but also with God. Therefore, business is not only aimed at achieving results but also maintaining the process as worship. Religious understandings and attitudes are one of the determinants of people's behavior that affects the consumption allowed by religion, namely, halal (Mukhtar & Mohsin Butt 2012).

Public awareness of halal products requires companies to provide products and services according to their religious beliefs in both the authentic and financial sectors (Thomas & Selimovic 2015). The call for companies to formally certify products and business practices and give them a halal label has become a new cultural phenomenon. This demand extends not only to countries with Muslim populations but also to non-Muslims throughout the world (Wilson 2014). This means that there is a shift in orientation from the halal domain to business, triggered by economic and technological reasons. Many companies are capturing this phenomenon as an opportunity by offering products and services through faith-based marketing (Izberk-Bilgin & Nakata 2016). On the other hand, this phenomenon raises concerns that companies ignore ethical values and ethical business objectives (Wilson 2014), both seen from deontological ethics, consequentialist ethics, and virtue ethics as in normative ethical theory (Jonsson 2011). This means that, in Islam, the authentic purpose of business is not only to pursue material gain but also spiritual with Islamic ethical principles, which are a synthesis of the three normative theories (Abdallah 2010).

---

*Corresponding author

DOI 10.1201/9781003189282-16

The study of the halal phenomenon has also become a new trend in academia with a variety of interesting themes such as consumer behavior (Amalia et al. 2020), cosmetic products (Briliana & Mursito 2017), microbial food ingredients (Karahalil 2020), industry and logistics (Ab Talib 2020), finance (Yusof et al. 2019) and so forth. In this study, our concentration on halal studies is in line with the development of the halal industry, which is growing very fast in the world (Hong et al. 2019). The development of the halal industry might be due to the increasing number of Muslims, which is even predicted to become the largest population in the world in 2060 (Lipka & Hacket 2017). This condition certainly affects the global economy and has the potential to form a promising market. Previous research conducted by Izberk-Bilgin and Nakata (2016) revealed that multinational companies such as Nestle', Unilever, McDonald's, Kentucky Fried Chicken, Campbell's, Colgate-Palmolive, HSBC, Tesco, Carrefour, and J. Walter Thompson (JWT) have targeted the Muslim market. They even call this segment of the Muslim community the next "one-billion" market after the Chinese and Indian markets. Other evidence also shows that more than a decade earlier, the world retail company, Walmart, opened a shop in Dearborn and Michigan to serve the large Muslim population in the area by offering halal meat, Islamic greeting cards, Islamic styles, and so on (Naughton 2008).

The literature has indicated that the potential of halal markets has penetrated religious and cultural boundaries. Albeit multinational companies are mostly owned by non-Muslims, their business activities have targeted the Muslim markets. This means that there is a shift from the domain of religious knowledge (halal) into business due to economic and technological factors. This concern has been shown by Wilson (2014) in one of his research conclusions that the values of Islamic business ethics will be ignored because of the focus on economic and business goals. Meanwhile, Muslim business people must apply business ethics in their trade, both for obtaining maximum profits and blessings. To achieve all of these intentions, the role of religion and culture is central to be applied in ethical and halal businesses. The understanding of business people about the laws in *fiqh al-muamalat* (known as business transactions according to Islamic law) is also imperative, both in terms of the laws of the product being transacted, the laws of the transaction process, and the goal of the halal guarantee (Muhammad 2020). In addition, the internalization of morals will guide efficient and ethical business whose success requires a strong commitment to Islamic ethics (Hassan 2016).

The study of halal with an impressive variety of themes shows that theoretical and empirical discussions have been carried out in various multinational businesses and modern markets. Nevertheless, specific studies on Islamic business ethics practices as a halal responsibility seem sparse, particularly in the local cultural practices based on religion in traditional markets that have specific uniqueness. To fill this void, the present study investigated the traditional market site of Tirtoyudo Malang, Indonesia, which adheres to the *"Pon"* market day and only lasts once in five days. Preliminary data shows that, when the *"Pon"* market arrived, the Tirtoyudo traditional market was filled with people from trade and economic, social, and cultural activities to the realm of tourist destinations. In addition, the community's obedience to the local culture and embodying of business and cultural activities in traditional markets has made Tirtoyudo the village chosen in the Village Building Movement (*Gema Desa*) in 2019 and has become an attraction for investors from the Indonesia-Japan Business Network.

## 2  METHODS

This study employed an exploratory descriptive design (Brink & Wood 1998) with a single case study (Yin 2014). It was carried out in 2020 within the traditional market of Tirtoyudo Malang, Indonesia, with the consideration of potential unique characteristics. Eleven participants consisting of head of market administrations (*Mantri*), traders, buyers, village leaders, and local governments, were recruited using a purposive and snowball sampling technique. Data was gathered from in-depth interviews with the participants (Yin 2014). The recorded and transcribed interviews were analyzed using an interactive model to construe the relationship of the category of activity variations in various cycles. The interactive analysis was carried out from the initial data collection process to the verification process for drawing conclusions (Miles et al. 2014). To maintain research quality standards, checking validity, comparing data, and verifying research data were necessary using the triangulation of data sources, methods, and other sources as well as member checks (Patton 2002).

## 3  RESULTS AND DISCUSSION

The Tirtoyudo traditional market was founded in 1965 by Tlogosari villagers, Tirtoyudo, Malang Regency, Indonesia. Located around 15 kilometers from the foot of Mount Merapi, which is inhabited by most Javanese and Madurese tribes and occupies an access area between the Java-Bali provinces through the southern route. Apart from being a place of trade, the Tirtoyudo traditional market is also a place a socio-cultural interaction. The traditional Tirtoyudo market plays an important role in increasing the economic value added of rural communities through ethical trade transactions and adherence to local culture. Our field exploration and interviews with the participants show that, in general, traders who carry out business transactions hold good ethical principles that are relevant to the teachings of Islam. These ethical principles are as follows.

### 3.1  *Trustworthy*

Most traders in the Tirtoyudo market are Muslim, although they still adhere to the principles of local culture passed down from their ancestors. Violation of the principles of local culture has severe social sanctions for people. Adherence to religion and the local culture has a positive influence on the business behavior of Tirtoyudo traditional market traders. Trustworthiness is an example. Trust is the basis of business transactions (Abuznaid 2009). The implementation of trustworthiness includes the positive correlation between actions and words, the conformity of reports and the reality of transactions and keeping promises. Based on our field exploration and interviews with the participants, the nature of this trust is revealed by the behavior of Tirtoyudo traditional market traders who convey product defects before sale or do not hide defects (*ghisysyi*), keep consumer or customer orders, do not exaggerate the quality of the goods sold, and offer goods in accordance with existing conditions. This was also admitted by one of the buyers who conveyed the same information. The results of the interview also show that the reason why this trustworthiness is important for traders is that they have an orientation so that their business results are better and halal.

Trustworthiness behavior in business starts from the registration of the traders. To become a member of a market, you must fill out a form, integrity pact and swear in front of the Market *Mantri* officers and community leaders not to commit fraud that harms others. Coaching in various local cultural events is also always carried out, for example, in "*kumpulan*" events (gathering between traders), "almsgiving or *bancaan*" (eating together as a form of gratitude) and informal gatherings. The Market *Mantri* and community leaders collaborate to become the central figures who regulate, guides and nurture the traders and ensures that the activities in the Tirtoyudo traditional market take place. Violation of the agreed rules has social and economic consequences that must be borne. For example, the perpetrator is announced in front of the general public, material fines are levied and they may not trade on the following market day. The purpose of this punishment is not out of hatred but solely so that no one is harmed.

The trustworthiness shown by Tirtoyudo traditional market traders is the practice of Sharia teachings that refer to the business success of the Prophet Muhammad SAW. The nature of mandate is later proven by modern economic and management theories about how a business should be run which aims to alleviate poverty through entrepreneurship in the paradigm of unity of God (*tauhid*) and divinity (Hunter 2014). Ethical principles such as the mandates practiced by Tirtoyudo traditional market traders are in line with the appreciation of Islam for business people who have an independent orientation, can maximize all their resources, and provide benefits to mankind. The function of trustworthiness in business helps people do good and avoid cheating and is in line with previous research on the implementation of Islamic ethics in business and its positive role (Jabbar et al. 2018).

## 3.2 *Fair*

Fairness is an important ethic in business and economic aspects. Our exploration and interviews show the fair nature of business. The fair implementation of business in the Tirtoyudo market is manifested by traders, among others, by not reducing the scales (*tahfif*) and cheating (*ghabn*). In addition, many business or store owners treat employees fairly regardless of race, ethnicity or class, and often even provide rewards based on the ethical performance of employees. In addition, the Market *Mantri* only gave permission for one person to sell (*bedak*) to make it evenly distributed. Traders are grouped based on the type of goods sold, making it easier for buyers. The price of basic necessities is determined as the same from one store to another, while the price of complementary goods varies according to the purchase price plus a reasonable margin.

Fair character values are formed through a long process. The rules agreed upon in the integrity pact are a trigger to act fairly and honestly. In addition, the role of the Market *Mantri* and village community leaders in campaigning for ethical behavior through local cultural events such as "*kumpulan*", "*bancaan*" and other events have a positive contribution. To ensure compliance with the values of fairness, the Market *Mantri* together with the team conducted monitoring by going around the market area to monitor the prices of basic necessities. This effort is made so that there is a similarity in the fair price limit offered to buyers. In addition, the Market *Mantri* randomly asked to the buyers about the price of the goods. Indirect monitoring is also carried out, through outreach, coaching together with the village community leaders and the provision of measuring instruments such as goods prices provided to the public. Buyers who feel aggrieved can take measurements through the measuring tools provided. Efforts were made by the Market *Mantri* with village community leaders in the framework of maintaining price stability, the condition of goods and achieving the ethical sustainability of the Tirtoyudo traditional market. The phenomenon of creating price stability through local culture strengthens the arguments of previous researchers who state that creating price stability through Sharia and ethical mechanisms is a very important aspect (Ahmed et al. 2018).

Fair values, such as the phenomenon in the Tirtoyudo traditional market, have a positive impact in increasing economic value-added and marketing local products. The success in maintaining ethical values also has an impact on the government award given to the Tirtoyudo village community through the Village Building Movement (*Gema Desa*) program in 2019 and has become an attraction for Indonesia-Japan Business Network investors. This traditional market activity has been going on for 55 years, apart from being one of the destinations for traders outside the Tirtoyudo village to market their commodities. It is also a marketing icon for local products, such as cocoa and coffee as well as a tourist destination that are in tune with the government's goals, namely, one village one product and one village one destination. This condition strengthens previous research stating that the core of marketing is the principle of value maximization based on justice (fair transactions, for example) to achieve the welfare of the wider community (Saeed et al. 2001).

The values of trustworthiness and fairness as practiced by Tirtoyudo market traders can also be seen from the perspective of traditional normative ethical theory. From the perspective of deontological ethics and consequentialist ethics, the values of trustworthiness and fairness are intrinsically true and have good consequences. The values of trustworthiness and fairness of traders that are practiced in their daily behavior and have become part of the work culture become individual virtues that come from the heart on the basis of strong faith. This condition by Jonsson (2011) is called virtue ethics, namely holistic virtue ethics that are not seen from just right and wrong but are used as life experiences within an ethical character. In the perspective of Islam, this is called *Ihsan*. A business that is run based on ethical values will ensure that every product produced is halal and will make a positive contribution to other lives. As villagers, they derive business ethics from empirical experiences (through local culture and religion) passed down from previous generations.

Table 1. Halal responsibility through Islamic business ethics practice.

| Islamic Ethics | Implementation of Islamic Ethical Values | Religious Foundation | Ethics Theory | Orientation |
|---|---|---|---|---|
| Trustworthy | • Positive correlation between actions and words<br>• Conformity of reports and the reality of transactions and keeping promises made<br>• Deliver defective products before sale or do not hide the defects (*ghisysyi*)<br>• Keeping customer or customer orders<br>• Do not overestimate the quality of the goods sold<br>• Offer goods in accordance with existing conditions | • Islamic Sharia<br>• Faith<br><br>• *Ihsan* | • Deontological Ethics<br>• Consequentialist Ethics<br><br>• Virtue Ethics | • Halal Responsibilities |
| Fair | • Not cheating on the scales (*tahfif*)<br>• Does not reduce the dose (*ghabn*)<br>• Treat employees fairly<br>• One person, one place to sell<br>• Traders are grouped based on the type of goods sold<br>• Prices of basic necessities are determined the same from one store to another | | | |

This empirical experience is then carried out with full awareness as a good religious adherent so that it becomes a culture. Abdallah (2010) calls this condition as the practice of a holistic Islamic ethical system which includes the obligation to carry out Islamic Sharia, based on faith with the motivation of Ihsan's virtues. The implementation of business ethics as a halal responsibility in trading in the Tirtoyudo traditional market is shown in Table 1.

## 4  CONCLUSION

The present study has uncovered the halal responsibility, which breaks through religious and cultural boundaries both in modern and traditional markets. An interesting phenomenon is captured in the Tirtoyudo traditional market, which is run by village people who have a halal responsibility by applying Islamic business ethics combined with adherence to local culture. Islamic business ethics has an important role in shaping businesses with Islamic character with trust and fairness principles in traditional markets. Achieving halal responsibility with the principle of trustworthiness and fairness through a coaching process by Market *Mantri* and community figures is supported by market entities on an ongoing basis in a blend of local cultures. The halal responsibility displayed in business practices in the Tirtoyudo traditional market fulfills a lifestyle and the consistent and ethical practice of *fiqh al-muamalat*. The goal is to achieve ethical economic development and be a bulwark to avoid the bad consequences of today's global market practises.

## REFERENCES

Ab Talib, M. S. 2020. Identifying halal logistics constraints in Brunei Darussalam. *Journal of Islamic Marketing, ahead-of-print* (ahead-of-print). https://doi.org/10.1108/JIMA-09-2019-0189

Abdallah, S. 2010. Islamic ethics: An exposition for resolving ICT ethical dilemmas. *Journal of Information, Communication and Ethics in Society*, 8(3), 289–301. https://doi.org/10.1108/14779961011071088

Abuznaid, S. A. 2009. Business ethics in Islam: The glaring gap in practice. *International Journal of Islamic and Middle Eastern Finance and Management*, 2(4), 278–288. https://doi.org/10.1108/17538390911100634

Ahmed, E. R., Islam, M. A., Alabdullah, T. T. Y., & bin Amran, A. 2018. Proposed the pricing model as an alternative Islamic benchmark. *Benchmarking: An International Journal*, 25(8), 2892–2912. https://doi.org/10.1108/BIJ-04-2017-0077

Amalia, F. A., Sosianika, A., & Suhartanto, D. 2020. Indonesian Millennials' Halal food purchasing: Merely a habit? *British Food Journal*, 122(4), 1185–1198. https://doi.org/10.1108/BFJ-10-2019-0748

Briliana, V., & Mursito, N. 2017. Exploring antecedents and consequences of Indonesian Muslim youths' attitude towards halal cosmetic products: A case study in Jakarta. *Asia Pacific Management Review*, 22(4), 176–184. https://doi.org/10.1016/j.apmrv.2017.07.012

Brink, P. J., & Wood, M. J. (Eds.). 1998. *Advanced design in nursing research* (2nd ed). Sage Publications.

El-Bassiouny, N. 2014. The one-billion-plus marginalization: Toward a scholarly understanding of Islamic consumers. *Journal of Business Research*, 67(2), 42–49. https://doi.org/10.1016/j.jbusres.2013.03.010

Hassan, A. 2016. Islamic ethical responsibilities for business and sustainable development. *Humanomics*, 32(1), 80–94. https://doi.org/10.1108/H-07-2015-0047

Hong, M., Sun, S., Beg, A. B. M. R., & Zhou, Z. 2019. Determinants of halal purchasing behaviour: Evidences from China. *Journal of Islamic Marketing*, 10(2), 410–425. https://doi.org/10.1108/JIMA- 03-2018-0053

Hunter, M. 2014. Entrepreneurship As A Means To Create Islamic Economy. *Economics, Management, and Financial Markets*, 9(1), 75–100.

Izberk-Bilgin, E., & Nakata, C. C. 2016. A new look at faith-based marketing: The global halal market. *Business Horizons*, 59(3), 285–292. https://doi.org/10.1016/j.bushor.2016.01.005

Jabbar, S. F. A., Ali, H. M., Mohamed, Z. M., & Jalil, F. 2018. Business Ethics: Theory and Practice in an Islamic Context. In J. Bian & K. T. Çalıyurt (Eds.), *Regulations and Applications of Ethics in Business Practice* (pp. 257–271). Springer Singapore. https://doi.org/10.1007/978-981-10-8062-3_14

Karahalil, E. 2020. Principles of halal-compliant fermentations: Microbial alternatives for the halal food industry. *Trends in Food Science & Technology*, 98, 1–9. https://doi.org/10.1016/j.tifs.2020.01.031

Lipka, M., & Hacket, C. 2017. *Why Muslims are the world's fastest-growing religious group | Pew Research Center*. https://www.pewresearch.org/fact-tank/2017/04/06/why-muslims-are-the-worlds- fastest-growing-religious-group/

Miles, M. B., Huberman, A. M., & Saldaña, J. 2014. *Qualitative data analysis: A methods sourcebook* (Third edition). SAGE Publications, Inc.

Muhammad, H. 2020. Holistic Practice of Fiqh Al-Muamalat: Halal Accountability of Islamic Microfinance Institutions. *Nusantara Halal Journal (Halal Awareness, Opinion, Research, and Initiative)*, 1(1), 22–31. https://doi.org/10.17977/um060.2020v1p022-031

Mukhtar, A., & Mohsin Butt, M. 2012. Intention to choose *Halal* products: The role of religiosity. *Journal of Islamic Marketing*, 3(2), 108–120. https://doi.org/10.1108/17590831211232519

Naughton, K. 2008. *Wal-Mart: Arab-America's Store*. Newsweek. https://www.newsweek.com/wal-mart- arab-americas-store-83885

Saeed, M., Ahmed, Z. U., & Mukhtar, S.-M. 2001. International Marketing Ethics from an Islamic Perspective: A Value-Maximization Approach. *Journal of Business Ethics*, 32(2), 127–142. https://doi.org/10.1023/A: 1010718817155

Thomas, P., & Selimovic, A. 2015. "Sharia on a Plate?" A critical discourse analysis of halal food in two Norwegian newspapers. *Journal of Islamic Marketing*, 6(3), 331–353. https://doi.org/10.1108/JIMA- 05-2014-0041

Wilson, J. A. J. 2014. The halal phenomenon: An extension or a new paradigm? *Social Business*, 4(3), 255–271. https://doi.org/10.1362/204440814X14103454934294

Yin, R. K. 2014. *Case study research: Design and methods* (Fifth edition). SAGE.

Yusof, R. M., Mahfudz, A. A., & Yaakub, S. 2019. Halal Trade Finance and Global Well-Being: Here Come the Millennials. In F. Hassan, I. Osman, E. S. Kassim, B. Haris, & R. Hassan (Eds.), *Contemporary Management and Science Issues in the Halal Industry* (pp. 469–494). Springer Singapore. https://doi.org/10.1007/978-981-13-2677-6_40

Halal Development: Trends, Opportunities and Challenges – Pratikto et al (eds)
© 2021 The Author(s), ISBN 978-1-032-03830-8

# Exploring small and medium industry stakeholders' perspectives of halal certification in Indonesia: A case of Bangkalan Regency, Indonesia

L. Qadariyah* & S. Nahidloh
*Universitas Trunojoyo, Madura, Indonesia*

A.I. Mawardi
*Universitas Islam Negeri Sunan Ampel, Surabaya, Indonesia*

ABSTRACT:   Halal certification is enacted to respond to the public anxiety of safe food products from forbidden materials based on Islamic rules. Anchored by small studies exploring halal certification perspectives among small and medium industry stakeholders in an Indonesian context, the present study was enacted to uncover these stakeholders' perspectives of halal certification. Fifteen small and medium industry owners (henceforth, participants) in Bangkalan, Madura, Indonesia, were involved in an interview. Findings suggest that the participants view halal certification as an essential aspect for selling their products. Also, due to the complicated process of achieving the halal certification label, most of their products are not certified.

*Keywords*:   halal certification, small and medium industry, perspective

## 1   INTRODUCTION

Small and medium industry (SMI) is a business unit that has an essential role in developing a community's economy. Besides being able to increase foreign exchange income for the country, this business unit can also help increase lower-class income. According to the Indonesian Ministry of Industry, it is explained that, in 2018, SMI could absorb a workforce of 17.9 million people and absorb 60% of the industrial workforce (Kemenperin 2019). These attainments have been seen as very significant in many countries, including Indonesia.

In Indonesia, SMI provides an ample contribution to national industrial development. In fact, 43.41% or 1.86 million are food sellers. At certain events, this percentage, especially in the food sector, increases due to the growing number of requests, such as before the *Idul Fitri* and other public holidays. Due to the large potential SMI brings, it is not surprising that the government also cares for them. Through capital assistance, routine coaching, and regular assistance, the government desires to develop the quality and quantity of SMI, including helping them in product halal certification. The halal status of a product has now become one of the considerations for consumers during the explosion of products that use non-halal ingredients, such as pork oil. It is explained in the Qur'an (Al-Maidah, 88):

> *O you who believe, do not prohibit what Allah has made legal for you, and do not transgress. Indeed, Allah does not like those who transgress. And eat food that is clean and good, from what Allah has blessed you with and fear Allah whom you believe in Him.*

Apart this verse, there are many other verses that instruct Muslims to pay attention to the halal and *thoyyib* consumption. According to al-Syatibi, Sharia law is principally aimed at realizing the

---

*Corresponding author

DOI 10.1201/9781003189282-17

benefit of mankind (Syatibi 2003). This means that all the commands and prohibitions of Allah to humans are for the good of humans themselves. When Allah commands people to consume halal food and *thoyyib*, there must be a reason for it whose purpose is to protect the human's life, soul, health, and others things.

This secret was then revealed through several studies that focused on discussing the positive effects of consuming halal food (Azeez 2013), including health and moral intelligence, psychology, and even a positive impact on human development in general (Arif & Ahmad 2011; Deuraseh 2014; Md. Sawari et al. 2015). Based on these advantages and also evidence of adherence to religious rules, the Institute for the Study of Food, Medicine, Cosmetics under the Indonesian Ulama Council enacted an initiative to certify or provide halal labels for food products that do not contain forbidden ingredients. This halal label, in its development, has become a trend not only in Indonesia but also abroad and is a potential market that is projected to continue to increase in the future (Elasrag 2016; Islam 2012).

This large potential is not only influenced by the increasing population of Muslims in the world (Global Islamic Economy Report 2019) but also the interest of non-Muslims to buy and consume halal products which for them are cleaner, more hygienic, and healthier (Atiah & Fatoni 2019; Hassan & Hamdan 2013; Mathew et al. 2014). With these considerations, the Indonesian Ulama Council is targeting that products circulating in Indonesia, whether from abroad or, especially domestic products must be certified starting October 2019. This policy continues to strive to meet these targets by working with mass organizations, universities, and several parties to become partners as a halal guarantee institution and which is also a mouthpiece for information on the importance of certification through education and mentoring programs. One of the targets of this program is small and medium industry sectors.

In the Bangkalan Regency, Indonesia, it was found that 130 SMI are in this area, according to a report by the Industry service. Most of the SMIs are engaged in the food sector, characterized by Madurese culture food. Unfortunately, most of them do not have halal certification. Therefore, the present study was designed to reveal how SMI stakeholders view halal certification. The findings are expected to shed light of the importance of intensive education and coaching for SMI stakeholders enacted by the Department of Industry and Trade.

## 2 METHODS

This study employed a qualitative approach that aims to describe holistically the small and medium industry stakeholders' perspectives of halal certification. Data was gathered through interviews with fifteen owners of this small and medium industry in Bangkalan, Madura, in the province of East Java, Indonesia. The data was analyzed using a reduction stage. We first carried out a data selection process, focusing on simplification, abstracting, and transforming data collected from written notes. Later, data presentation was carried out in the form of data preparation for drawing conclusions.

## 3 RESULTS AND DISCUSSION

### 3.1 *Legal basis for halal certification*

The legal basis for the enforcement of halal certification is derived from Sharia law (*al-hukmasy-syar'i*). To ensure the enforcement of these Sharia provisions in relation to the law of halal and haram, procedural regulations (*al-hukm asy-syar'i*) are needed. The legal basis for the validity of halal certification is as follows:

> Then eat what is lawful and good and the sustenance that Allah has given you, and be thankful for Allah's blessings, if you only worship Him (Surah An-Nahl [16]: 114).

In connection with the recommendation to eat and drink lawful food or beverage, several verses have informed these, including *Surah al Baqarah*, 168, and *Nahl* 114, *al Midah* 88. Consumption in Islam is limited by Sharia rules, where there are several things that are halal and haram for consumption (Havis 2017). For the Islamic society, consuming what is lawful and good (*thayyib*) is a manifestation of devotion to Allah. Muslim consumers' behaviour is different from non-Muslim consumers. Muslim consumers, both in terms of production and consumption, are bound by Islamic rules, meaning that the freedom related to the production and consumption activities is regulated. Since Islam prohibits a Muslim from consuming several commodities, such as alcohol and pork, then even in the production of food it is prohibited to use these ingredients (Al-Teinaz et al. 2019; Alzeer et al. 2018). Therefore, the existence of a halal label guarantees that the food is protected from things that are prohibited in Islam.

### 3.2 Halal certification in the perspective of small and medium industry (SMI) stakeholders in Bangkalan Regency, Indonesia

As mandated by Law No. 33 of 2014, in 2019, all halal products spread across Indonesia, both local and foreign products, are required to have a halal label. This mandate is enacted by the government so that the domestic industry can compete in the world halal industry which is currently a potential market. Therefore, it is not surprising that the government, through several related agencies, makes such efforts to be able to teach the importance of the halal label and also provide intensive guidance to them regarding the halal labelling process.

In Bangkalan regency, SMI under the governance of the Department of Industry and Trade have received trainings on halal certification and labelling, despite efforts to the contrary. In terms of halal certification and labelling, food is quite different from other products. In this regard, in order to obtain a halal certificate, the raw materials, production process, expiration period, and nutrition are required and objectively observed. Therefore, SMI that have accomplished these processes is given halal certification and labialization as formal proof that the products sold are well-guaranteed for halal consumption.

Our interview with fifteen SMI owners shows that the participants view halal certification as an essential aspect for product selling and consumption. They believe that products with a halal label can be ascertained to be of good quality and are safe to consume. This perspective is seen as a picture of Madurese religious and cultural characteristics. These cultural specificities are seen in their hierarchical obedience, submission, and surrender to the four main figures such as father (*bhup-pa'*), mother (*bhabbu'*), teachers (*ghuruh*) and government (*ratoh*) (Wiyata 2020). In religious practice, Madurese society strongly adheres to the teachings of their religion. Islamic symbols are central and become a serious concern, including the need for a halal label.

We also discovered that 33.3% of the participants in this study are not consistent in checking the halal label on the materials to be used in the products by arguing that they do not need to check as the products are natural ingredients such as herbal medicine. It can be concluded that they consider it unimportant to check halal materials in the products since the materials such as cooking oil and sugar are their basic ingredients for the products. For instance, regardless of the products, sugar is taken from sugar cane and cooking oil from coconut which naturally makes these materials halal.

The participants' understanding on these natural halal materials are derived from Madurese cultures (Wiyata 2020). they will do anything according to their beliefs and knowledge. When they believe that an item is halal, even if there is no halal logo, it will be considered lawful for consumption. In a similar vein, Madurese people are consistent in upholding their religious principles, including for haram product consumption (Jonge 1989; Rifai 2007; Subaharianto 2004).

Based on Madurese understanding, halal food does not contain *khamr*, pork, and others. The aforementioned understanding refers to a global material which excludes pork oil and skin, as explained in *Surah al Maidah*, 3:

> It is forbidden for you to (eat) carcasses, blood, pork, (animal meat) that were slaughtered in the name other than Allah, those who were strangled, those who were beaten, those who fell, who were gored, and were killed by wild animals, except those you had slaughtered them, and (forbidden). for you) who were slaughtered for idols (al Maidah: 3).

In the context of the halal label, the participants also viewed it as an important element in convincing consumers, albeit some of the participants rarely check for the halal label. It indicates that the halal label has been a requirement for selling halal-guaranteed products and can help consumers to avoid deception. When examined from the theory of producer behavior in Islam, producers in Islam are required to implement moral values following the Qur'an and *Hadith* in terms of meeting consumers' needs, production processes, obtaining capital, business growth, and diversification for business continuity (Sukarno 2010).

Some of the participants in this study argued that a halal label can also be used to increase their sales quantity as previous research has shown (Hanim Yusuf et al. 2016). These SMI owners also have a good perception of the relationship between halal labels and increased sales. Most of the participants (13 out of 15) believed that this halal label affects the level of sales, since with this halal label, the product is well accepted in the middle of society. When asked whether they had the desire to register their products in order to obtain halal certificates and labels, 60% stated that they had the desire to register their products as products with a halal label, 6.7% stated that they already had a halal label on their products, and equal to 20% stated that they were in the process of certification and labelling and 13% did not answer these questions. Most of the reasons why they do not have halal certificates until now are because of the cost which is considered quite expensive and the process is quite long. Our findings should use as catalysts for the Indonesian government to enact society-based policies, especially in the halal certification sector.

## 4 CONCLUSION

This study has shown that, from the perspectives of small and medium industry stakeholders, halal certification is a central aspect for traded products. This label is quite influential in increasing sales circulation. However, despite the halal label significance in a product, very few of the participants involved in this study checked the product halal ingredients. In other words, their understanding of halal certification is seen in this study, but their sustained attention to halal ingredients is not held.

## REFERENCES

Al-Teinaz, Y. R., Spear, S., & Abd El-Rahim, I. H. A. (Eds.). 2019. *The halal food handbook*. John Wiley & Sons, Inc.

Alzeer, J., Rieder, U., & Hadeed, K. A. 2018. Rational and practical aspects of Halal and Tayyib in the context of food safety. *Trends in Food Science & Technology, 71*, 264–267. https://doi.org/10.1016/j.tifs.2017.10.020.

Arif, S., & Ahmad, R. 2011. Food quality standards in developing quality human capital: An Islamic perspective. *Afican Journal of Business Management, 5* (31).

Atiah, I. N., & Fatoni, A. 2019. Sistem Jaminan Halal: Studi Komparatif Indonesia dan Malaysia. *Syi'ar Iqtishadi: Journal of Islamic Economics, Finance and Banking, 3*(2), 37. https://doi.org/10.35448/ jiec. v3i2.6585.

Azeez, W. 2013. The Halal Dietary System as a Recipe for Good Health. *IOSR Journal of Humanities and Social Science, 7*(4), 7.

Deuraseh, N. 2014. Islamic Dietary Habit for Preservation of Health: A Review from Islamic Literature. *Middle East Journal of Scientific Research, 20 (2)*, 7. https://www.researchgate.net/pro-file/Nurdeng_Deuraseh/ publication/305298866_Islamic_Dietary_Habit_for_Preserva-tion_of_Health_A_Review_from_ Islamic_ Literature/links/57872e2608aef321de2c7a15/Islamic-Die-tary-Habit-for-Preservation-of-Health-A- Review-from-Islamic-Literature.pdf.

Elasrag, H. 2016. Halal Industry: Key Challenges and Opportunities. *SSRN Electronic Journal.* https://doi. org/10.2139/ssrn.2735417. *Global islamic Economy Report. 2019.*

Hanim Yusuf, A., Abdul Shukor, S., & Salwa Ahmad Bustamam, U. 2016. Halal Certification vs Business Growth of Food Industry in Malaysia. *Journal of Economics, Business and Management, 4*(3), 247–251. https://doi.org/10.7763/JOEBM.2016.V4.399.

Hassan, S. H., & Hamdan, H. 2013. Experience of Non-Muslim Consumers on Halal as Third Party Certification Mark in Malaysia. *Asian Social Science, 9*(15), p263. https://doi.org/10.5539/ass.v9n15p263.

Havis, A. 2017. *Sejarah Pemikiran Ekonomi Islam Kontemporer*. PT. Kharisma Putra Utama.

Islam, T. 2012. Halal Marketing: Growing Pie. *International Journal of Management Research and Review*, *3*(12), 11.

Jonge, H. D. 1989. *Madura dalam Empat Zaman: Pedagang, Perkembangan, Ekonomi dan Islam: Suatu studi Antropologi Ekonomi*. Gramedia.

Kemenperin. 2019. IKM Berkontribusi 60 Persen Serapan Total Tenaga Kerja Industri. *IKM Berkontribusi 60 Persen Serapan Total Tenaga Kerja Industri*.

Mathew, V. N., Abdullah, A. M. R. binti A., & Ismail, S. N. binti M. 2014. Acceptance on Halal Food among Non-Muslim Consumers. *Procedia – Social and Behavioral Sciences*, *121*, 262–271. https://doi.org/10.1016/j.sbspro.2014.01.1127.

Md. Sawari, S. S., Ghazali, M. A., Ibrahim, M. B., & Mustapha, N. I. 2015. Evidence Based Review on the Effect of Islamic Dietary Law Towards Human Development. *Mediterranean Journal of Social Sciences*. https://doi.org/10.5901/mjss.2015.v6n3s2p136.

Rifai, A. M. 2007. *Manusia Madura: Pembawaan, Perilaku, Etos Kerja, Penampilan dan Pandangan Hidupnya Seperti dicitrakan Peribahasanya*. Pilar Media.

Subaharianto, A. 2004. *Tantangan Industrialisasi Madura: Membentuk Kultur, Menjunjung Leluhu*. Bayumedia.

Sukarno, F. 2010. Etika Produksi Perspektif Ekonomi Islam. *Jurnal Ekonomi Islam al Infaq*, *1*(1), 13.

Syatibi, A. I. al. 2003. *Al Muwafaqat fi Usul al Syariah Jilid I Cet III*. Dar Kutub al Ilmiyah.

Wiyata, L. 2020. *Carok; Konflik Kekerasan dan Harga Diri Orang Madura*. LKiS.

# The influence of demographics, socio-economics, and the environment on the preferences and behavior of middle-class Muslims in forming the potential for a halal hospital

L. Latifah*, Fatmah, A.I. Mawardi & I. Ritonga
*Universitas Islam Negeri Sunan Ampel, Surabaya, Indonesia*

ABSTRACT: This study was designed to unveil the influence of demographic, socio-economic, and environmental factors on the preferences and behavior of middle-class Muslims toward the potential of a halal hospital in Surabaya, Indonesia. A simple random sampling technique was used with a total of 217 respondents in this study. To analyze the data, the Structural Equational Model (SEM) from the AMOS statistical software was used. Findings revealed that demographics, socio-economics, and the environment significantly influenced respondents' preferences. This study also discovered that preferences towards behavior, preferences towards the potential of a halal hospital, and the actual behavior towards the potential of a halal hospital are correlated. Arguably, a halal hospital's potential is central, which is around 75% as measured by the level of preference and behavior of middle-class Muslims who have different demographic, social, economic, and environmental characteristics.

*Keywords*: Middle-class Muslim, demographic, socio-economic, environment, potential halal hospital

## 1 INTRODUCTION

Middle-class consumers have been emerged as a marketing revolution in Indonesia in the last five years. According to the Central Statistics Agency, the increase in the middle-class population with expenditure per day of 2-20 USD in Indonesia has now reached around 8-9 million people annually (Central Statistics Agency 2013), so that the number of Indonesia's middle-class population has reached 130 million. This phenomenon has significant market potential as this population has more income. The potential of middle-class consumers in Indonesia, especially Muslim consumers, can be seen from the rise of Muslim middle-class markets in Islamic banking with a growth of 40% since 1991. Looking at this advancement, it is thus central to consider the principle of halal products or services, including in health service sectors.

Surabaya, the capital city of East Java province in Indonesia, is the second-largest city in Indonesia after Jakarta, with a population of 3,016,344 people. This percentage includes the middle class Muslim population of 2,489,690 people (82.54%). The high number of middle-class Muslims in Surabaya allows us to find out how their preferences and behavior would affect halal hospital services and health care. The investigation of their preferences and behaviors is essential since most of them consume halal products in the city. Preference in this study includes choices that consumers prefer. A preference ranks all situations or conditions starting from the most preferred to the least preferred. Consumer preference is the nature of the desire of consumers to choose existing products or services. Preferences in the Islamic perspective are also well explained. According to Madrasir and Khoiruddin (2012), several principles should be aligned with the Islamic perspective, such as

---

*Corresponding author

DOI 10.1201/9781003189282-18

(1) the excellence goods and services, (2) the usefulness or use of goods and services consumed, (3) the appropriate quantity of goods and services consumed.

Behavior is an action or deed of an organization that can be observed and can even be learned. Consumer behavior is an action involved in the acquisition and consumption of products ad services, including the processes that precede and follow consumption (Husein 2005). Understanding consumer behavior more accurately will make it easier for manufacturers to prepare the right marketing techniques to deal with consumer behavior. According to Etta and Sofia (2013), three basic concepts of consumer behavior include (1) the need for some basic satisfaction or desire to satisfy specific needs, (2) the demand for a specific product supported by the ability and willingness to buy it.

In general, there are two approaches to defining the middle class: the relative approach and the absolute approach (Easterly 2001; Yuswohadi 2014). One of them is used by Asia Development Bank, which defines the middle class as people with a per capita expenditure range of 2-20 USD, which is divided into lower middle classes (2-4 USD), middle classes (4-10 USD), and upper middle class (10-20 USD). A Sharia hospital (also known as a "halal hospital") has been outlined by the fatwa of the National Sharia Council of the Indonesian Ulema Council No. 107/DSN-MUI/X/2016 with guidelines for Hospital Administrators to use based on Sharia Principles. The public may need an explanation of the guidelines for the organization of hospitals based on Sharia principles and legal provisions.

The fatwa regarding Sharia hospitals is the basis for analyzing its potential in the present study, including (1) the provisions related to contracts and legal issues, (2) the provisions related to hospital services, (3) the provisions related to the use of drugs, food-beverages, cosmetics, and used goods, (4) the provisions related to the placement, use, and development of hospital funds, and (5) the provisions related to safeguarding religion, the soul, intellect, heredity, and treasure as explained in *maqashid al-Shariah* or the purpose of Sharia enactment. This study sought to investigate: (1) the influence of demographic, social, economic, and environmental variables on the preferences of middle-class Muslims in shaping the potential of a Halal Hospital, (2) the preferences' effect on the behavior, and (3) the effect of the preferences and behavior on the potential of Halal Hospital.

## 2 METHODS

Garnered through Structural Equation Modeling (SEM) (Jogiyanto 2011) with AMOS statistical programs, the study sites were based in Surabaya City areas covering West Surabaya, South Surabaya, North Surabaya, Central Surabaya, and East Surabaya. "Middle-Class Muslims" (hence-forth, "respondents") includes 3,016,344 people, with 82.54% or 2,489,690 people with expenditures above 2 USD per day. Also, 2,117,482 (85.05%) respondents are Muslim residents. We recruited 217 respondents using a simple random sampling technique. The analysis was carried out after previously measuring the model using validity and reliability tests. Conceptually, If the construct is unidimensional, it is analyzed using first-order confirmatory factor analysis. In contrast, if it is multidimensional, it is analyzed using a second-order confirmatory factor (Notoatmodjo 2012).

## 3 RESULTS AND DISCUSSION

This study documents that the composition of the middle-class Muslims' age in Surabaya is 140 adults 26-45 years of age in a total of 217 respondents, with as many as teenagers 20.3% of the sample and elderly people representing 15.2%. Meanwhile, the factor of seniority or the emergence of the older group is more dominant in channeling preferences, including in making decisions about their health. It is natural that this age group is less dominant in giving preference, including making decisions. This is in accordance with what Müller-Arteaga et al. (2020) has explained: that there is a difference of position based on age which will change the old and young groups who are different in terms of channeling preferences and making decisions. According to Wiyono (2006), age influences a person's active participation.

106

In the context of the demographic factor, the composition between women and men is not much different or almost the same, with 104 men (47.9%) and 113 women (52%) out of 217 respondents. Although the difference is small, this gender difference is significant in influencing the decision or preference. This finding is consistent with SEM model measurements that have a positive loading factor of 1.00. Theoretically, gender influences the desire and ability of the community, especially the middle-class Muslims, to participate in a community. According to Mappiare (1994), men and women differ in their perspectives in dealing with a problem. Thus, the Male group has the privilege compared to the female group. Thus, male groups have more opportunities to participate.

In the context of socio-economic factors, most Muslims in Surabaya have attained tertiary education (83.4%), and 15% of them educated are less educated. These levels of education affect the changes in attitudes and behavior in healthy living. A higher level of education enables people to absorb information and implement it in their daily behavior and lifestyle, especially in the case of halal hospitals. Economic factors in this study are explained by two aspects, namely the work variable and respondent expenditure variable. The Job variable or type of business is the work occupied by respondents or the middle-class community in Surabaya. The expenditure variable is the amount of respondents' expenditure for the needs of respondents (Kotler 2007).

According to the Indonesian Central Statistics Agency, the average expenditure per capita is the costs incurred for consumption per household member for a month, both from gifts, purchases, and income. The average expenditure per capita of Surabaya's middle-class Muslim community is included in the lower-middle-class group, which is 75.6% or 164 people out of 217 total respondents surveyed. The lower middle class's high level will economically increase the number of products and services, including health services such as halal hospitals as there will be more and more Muslim middle-class money to be spent (Zarmani et al. 2015).

In the context of the environment and hospital facilities, 80% of the respondents are neutral regarding health services that have been visited, both functionally, emotionally, and spiritually with 5.709 and CR 1.780 (>0.5), and a loading factor of 10.164 (>0.7). It implies that the work variable is very large. It can be seen that 22.6% of the respondents or around 49 people out of 217 total respondents are self-employed, then 19.4% are traders, 17.1% are private employees, 13.8% run businesses in the services, 9.2% have their own production, 0.5% are in agriculture and fisheries, and only 3.2% are serving as civil servants, military, and police (Rosyada 2015; Veithzal 2014).

Our study also portrayed the middle-class Muslims' preference variable on the halal hospital potential. The results showed there was a significant influence between the preferences and behavior of the Muslim middle class on the potential of Halal Hospital in Surabaya. As explained by Kotler (2010), in his book Marketing Management, preferences mean preferences, choices, or something that consumers prefer. Preferences in this study were documented from Muslims in Surabaya regarding a halal hospital that is in accordance with Sharia principles. The organization and service of a hospital that is in accordance with the principles of Sharia are in accordance with the fatwa of DSN MUI (2018) No.107 concerning the implementation of Sharia hospitals in which there are legal bases in accordance with the Qur'an and Al-Hadith and the principles of *fiqh*. The background is that there must be a fatwa about Sharia hospitals.

The basic principles of Sharia that form the basis of research in shaping the preferences of middle-class Muslim communities in the city of Surabaya in this study are the existence of legal agreements and personnel, hospital services, payment transactions, the use of medicines and food and drinks, cosmetics and used goods and *maqashid al-syariah*, namely protecting religion, reason, life, descent and wealth. This preference was built by taking into account the demographic, social, economic, and environmental factors of the middle-class Muslim community of Surabaya City.

Middle-class Muslims in this study are demographically mature in age with a per capita expenditure level mostly in the lower middle class, mostly with high education and have a universalist lifestyle. As explained by Yuswohadi (2014) in his book Marketing to the Middle Class Muslim, which divides the lifestyle of middle class Muslims into 4, namely apathis, rationalist, conformist and universalist. In this study, the universalist lifestyle was 37%, the apathetic lifestyle was 35%, the rationalist lifestyle was 19% the conformist lifestyle was 9%.

The universalist lifestyle is the lifestyle of the largest middle class in Surabaya with the character-istics of tolerance, open-mindedness, upholding universal values and inclusive of values outside of Islam. From the characteristics of Surabaya's Muslim middle-class consumers, it can be concluded that Surabaya's Muslim middle-class consumers are Muslim consumers who have broad knowledge or insight into global mindsets and have technological literacy and, on the other hand, firmly carry out Islamic values in everyday life. They understand and apply Islamic values substantively, not just normatively. They are more willing to accept differences and tend to uphold universal values, they are not ashamed to be different, and they tend to accept the differences of others.

We concluded that the strength and ability of the middle-class Muslims in Surabaya is very large in relation to realizing it is halal hospital. By understanding middle-class Muslim preferences in Surabaya, investors or policymakers will be able to design appropriate strategies to respond to consumer expectations. In the SEM statistical test, the effect of preference on potential was SE: 0.021, CR 4.325, and a loading factor of 0.092. This means that the construct of preferences as measured by demographic, social, economic, and environmental variables significantly influences the potential for halal hospitals. This means that the constructs of the preferences of middle-class Muslims who have different demographic, social, economic, and environmental backgrounds are related to halal hospitals or hospitals based on Islamic Sharia principles that are related to legal contracts and personnel, home services sickness, payment transactions, the use of drugs and food and drinks, cosmetics and used goods, the protection of religion, the guarding of reason, the protection of souls, the guarding of children and the protection of possessions have very high potential.

## 4 CONCLUSION

This study has revealed that demographic, social, economic, and environmental variables have a significant influence on the preferences of middle-class Muslims in Surabaya toward the halal hospital. This is indicated by the demographic value of SE: 5.433 CR: 0.573, social value of SE: 2,473 CR: 0.787, economic value of SE: 4,032 CR: $-1,338$, and environmental value of SE: 2,111 CR: 0.0558. Also, this study uncovered that middle-class Muslims' preferences in Surabaya have a significant influence on their behavior. This is indicated by the value of SE: 0.024 CR: 7,614. More importantly, their preferences and behavior significantly influence the potential of the halal hospital. This is indicated by the preference variable with a value of SE: 0.083 CR: 0.774 and P: 0.000 and the behavioral variables value of SE: 0.083, CR: 7.774, and P: 0.000.

## REFERENCES

Central Statistics Agency. 2013. *Profile kesehatan Indonesia.* Kemenkes Republik Indonesia.
DSN-MUI. 2018. *Pedoman Penyelenggara RumahSakit Berdasarkan Prinsip Syariah*, https://dsnmui.or.id.
Easterly, William. 2001. *The Middle-Class Consensus and Econommic Development,* Journal of Economic Growth, vol. 6. Swedia: EconPapers Orebro University.
Husein, Umar. 2005. *Riset Pemasaran dan Perilaku Konsumen,* Jakarta: PT Gramedia Pustaka Utama dan. Jakarta Busnis Researct Center.
Jogiyanto. 2011. *Konsep dan Aplikasi Struktural Eqution Modeling,* Yogyakarta: UPPSTIM YKPN.
Kotler, Philip. 2000. *Manajemen Pemasaran,* Jakarta: Prehalindo.
Kotler, Philip, Keller L K. 2007. *Manajemen Pemasaran*, Jakarta: Pearson Education Inc.
Madrasir and Khoirudin. 2012. *Etika Bisnis Dalam Islam,* Lampung: Seksi Penerbitan Fakultas Syariah IAIN Raden Intan.
Mappiare, Andi. 1994. *Psikologi Orang Dewasa Bagi Penyesuaian Dan Pendidikan,* Surabaya: Usana Off-set printing.
Müller-Arteaga, C., Martín Martinez, A., Padilla-Fernández, B., Blasco Hernández, P., Espuña Pons, M., Cruz, F. & López-Fando, L. 2020. Position of the Ibero-American Society of Neurourology and Urogynecology in relation to the use of synthetic suburethral meshes for the surgical treatment of female stress incontinence. *Neurourology and Urodynamics, 39*(1), 464–469.

Notoatmodjo, Soekidjo. 2012. *Metodologi Penelitian Kesehatan.* Jakarta. Rineka Cipta.

Rosyada, Dede. 2015. Seminar on *Rumah Sakit Syari'ah 2015*, FKIK UIN Syarif Hidayatullah Jakarta.

Veithzal. 2014. *Manajemen Sumber Daya Insani.* Jakarta. PT Raja Grafindo Persada.

Wiyono, Slamet, 2006. *Manajemen Potensi Diri*, Jakarta: PT Grasindo.

Yuswohadi. 2014. *Marketing to the Middle-Class Muslim,* Jakarta: PT Gramedia Pustaka Utama.

Zarmani, Nur Farhani, M. Anuar, Shaikh M. Saifuddeen. 2015. *Potensi Pemba ngunan Industri Peranti Halal Terhadap Penjanaan Sektor Pelancongan Perubahan Halal di Malaysia.* Proceedings of the 2nd International convention on Islamic Management.

# Author index